Hiwa Asadpour and Thomas Jügel (Eds.)
Word Order Variation

Studia Typologica

Beihefte / Supplements
STUF – Sprachtypologie und Universalienforschung
　　　Language Typology and Universals

Editors
Thomas Stolz, François Jacquesson, Pieter C. Muysken

Editorial Board
Michael Cysouw (Marburg), Ray Fabri (Malta), Steven Roger Fischer (Auckland), Bernhard Hurch (Graz), Bernd Kortmann (Freiburg), Nicole Nau (Poznán), Ignazio Putzu (Cagliari), Stavros Skopeteas (Göttingen), Johan van der Auwera (Antwerpen), Elisabeth Verhoeven (Berlin), Ljuba Veselinova (Stockholm)

Volume 31

Word Order Variation

Semitic, Turkic and Indo-European Languages in Contact

Edited by
Hiwa Asadpour and Thomas Jügel

DE GRUYTER
MOUTON

ISBN 978-3-11-163181-3
e-ISBN (PDF) 978-3-11-079036-8
e-ISBN (EPUB) 978-3-11-079053-5
ISSN 1617-2957

Library of Congress Control Number: 2022934663

Bibliographic information published by the Deutsche Nationalbibliothek
The Deutsche Nationalbibliothek lists this publication in the Deutsche Nationalbibliografie;
detailed bibliographic data are available on the Internet at http://dnb.dnb.de.

© 2024 Walter de Gruyter GmbH, Berlin/Boston
This volume is text- and page-identical with the hardback published in 2022.
Cover image: Alpha-C/iStock/Thinkstock

www.degruyter.com

Preface

The idea for this volume was born on occasion of the panel "Crosslinguistic Study of Word Order in Western Asia: A Diachronic-Synchronic Perspective", which was organized by Hiwa Asadpour at the international conference "Anatolia-the Caucasus-Iran: Ethnic and Linguistic Contacts (ACIC)", held from 10–12 May 2018 in Yerevan, Armenia.[1]

Word order in languages of the Semitic-Turkic-Iranic contact area received special attention thanks to seminal work by Haig (notably 2015, 2017, to appear). Iranic languages are said to be SOV (subject-object-verb) languages, but a number of them display constituents after the verb, notably Goals. Haig considered this an areal phenomenon and coined the term "Mesopotamian OVG word order" (Haig 2017: 410). This is accentuated in Haig and Khan (2019: 24) who identify "northern Iraq and neighbouring regions of western Iran" as the "areal epicentre" for OVG word order, which suggests that this was its area of origin.

The present volume examines these statements by providing various studies on languages from inside and outside the contact zone, which investigate word order variation with a focus on objects and Targets (T) in particular. "Target" is used here as an umbrella term covering various semantic roles which represent the endpoint of expressions of virtual or figurative motion (Asadpour 2021). These usually include Goals, Recipients, Addressees, Themes of verbs such as "look", and results of Change-of-State verbs. Although Beneficiaries overlap to some extent with Targets, Asadpour (2021, §9.1) shows that they are not prototypical representatives of this class.

Contact-induced word order change is evidenced by a number of contributions in this volume (Bulut, Herin, Lemskaya, Neocleous, Noorlander and Molin). However, the peculiar OVT order is also attested outside the contact zone and in historical data (Jügel, Korn). Many highlight the relevance of grammatical, discourse-pragmatic, and cognitive principles in order to account for the complex situation of word order variation (in particular Asadpour, Payne, Wasow). As such, language contact may support already ongoing development or tip the scales in favor of one of several existing patterns.

The volume does not work within a narrow theoretical framework, but rather explores many different empirical dimensions and theoretical approaches

---

[1] The panel was part of the "International Cooperation on Contact Linguistics in Cross-Border Kurdistan", funded by the German Research Foundation (DFG).

(morpho-syntactic, cognitive, discourse-pragmatic) and thereby presents a whole raft of cogent evidence for empirical and theoretical conclusions.

Frankfurt/Bochum, Hiwa Asadpour and Thomas Jügel, January 2022

References

Asadpour, Hiwa. 2021. *Typologizing word order variation in Northwestern Iran*. Frankfurt: Goethe University Frankfurt PhD Thesis.

Haig, Geoffrey. 2015. Verb-Goal (VG) word order in Kurdish and Neo-Aramaic: Typological and areal considerations. In Geoffrey Khan & Lidia Napiorkowska (eds.), *Neo-Aramaic and its linguistic context*, 407–425. Piscataway: Gorgias Press.

Haig, Geoffrey. 2017. Western Asia: East Anatolia as a transition zone. In Raymond Hickey (ed.), *The Cambridge handbook of areal linguistics*, 396–423. Cambridge: Cambridge University Press.

Haig, Geoffrey. to appear. Post-predicate constituents in Kurdish. In Yaron Matras, Ergin Öpengin & Geoffrey Haig (eds.), *Structural and typological variation in the dialects of Kurdish*. London: Palgrave MacMillan.

Haig, Geoffrey & Geoffrey Khan. 2019. Introduction. In Geoffrey Haig & Geoffrey Khan (eds.), *The Languages and linguistics of Western Asia – An areal perspective*, 1–29. Berlin & Boston: De Gruyter Mouton.

Contents

Preface —— V

Thomas Wasow
Factors influencing word ordering —— 1

Doris L. Payne
Proximal to distal: Information flow and order in Maa —— 15

Thomas Jügel
Word order variation in Middle Iranic: Persian, Parthian, Bactrian, and Sogdian —— 39

Hiwa Asadpour
Word order in Mukri Kurdish – the case of incorporated Targets —— 63

Agnes Korn
Targets and other postverbal arguments in Southern Balochi: A multidimensional cline —— 89

Bruno Herin
Word order in contact and the expression of Target in Northern Domari —— 127

Nicolaos Neocleous
The evolution of VO and OV alternation in Romeyka —— 141

Christiane Bulut
Word order in Iran-Turkic —— 163

Valeriya Lemskaya
Word order variation in Chulym Turkic of Siberia —— 181

Daniel Birnstiel
Copulas and Target phrase positioning in the Arabic dialects of Kurdistan. Preliminary remarks on two information-structure related features —— 197

Paul M. Noorlander and Dorota Molin
Word order typology in North-Eastern Neo-Aramaic —— 235

Index of Authors —— 259

Index of Languages —— 262

Index of Subjects —— 264

Thomas Wasow
Factors influencing word ordering

Abstract: A number of factors have been shown to influence word order preferences, both across languages and within particular languages. These include preferences for: subject before object; verb adjacent to direct object; short phrases before long phrases or the reverse, depending on the headedness pattern of the language; old information before new information; and more accessible before less accessible information. Plausible explanations for many of such word order tendencies have been offered in terms of ease of processing, discourse coherence, or nuances of meaning to be expressed.

Keywords: accessibility; communication; headedness, information structure; weight

1 Introduction

Speech unfolds over time, necessitating that words in spoken languages appear in some sequential order. In principle, that ordering could be free; that is, it is logically possible that word order plays no role in the well-formedness or meaning of utterances. But that is not the case. Every natural language seems to have some constraints on possible word orders. Even in so-called free word order languages, some orderings are ungrammatical, and differences among the grammatical orders can serve communicative functions (see, for example, Simpson 2007).

Constraints on word ordering are to some extent conventionalized, as is evident from the fact that different languages have different patterns of word order. But there are some generalizations that hold across languages, both in terms of the dominant patterns of word order and in terms of the types of word order variation we find within and across languages.

The word order generalizations discussed in this chapter are all probabilistic – that is, they have exceptions. There are no doubt some restrictions on word order in particular languages that hold without exception. I do not, however, know of any exceptionless universal constraints on word order.

Thomas Wasow, Department of Linguistics, Stanford University, Stanford, USA,
E-mail: wasow@stanford.edu

A great deal of the literature on word order has been primarily descriptive. Two of the pioneers in this domain, Otto Behaghel and Joseph Greenberg, discovered some of the most important generalizations about word order but wrote little about why those generalizations hold. Starting with Hawkins (1994), however, much of the research on word order has looked for explanations as to why languages exhibit the word orders they do. The explanations are of two general types, which I will refer to as processing and pragmatic.[1] Processing explanations account for word order preferences on the basis of communicative efficiency. That is, some word orders are more easily produced or comprehended than others, owing to either general properties of human cognition or to even more general properties of information transmission. Pragmatic explanations account for word order preferences by showing that they convey aspects of meaning that go beyond literal meaning (that is, truth conditions); these include conveying the relative importance of parts of a sentence or how the sentences in a discourse are connected.

This chapter reviews some of the main generalizations about word order discovered by Behaghel, Greenberg, and others, and discusses some of the work that has been done to try to explain those generalizations. This is a rich literature, and I do not pretend to be able to provide a comprehensive survey; rather, the chapter is a sampling of research on this topic, along with some reflections on the research described. For a more thorough survey, see Song (2012).

2 Cross-language patterns

Greenberg's seminal 1963 paper "Some universals of grammar with particular reference to the order of meaningful elements" was the first of a very large number of typological studies of word order. One of its important contributions was the classification of languages on the basis of the relative positions of verb, subject, and object in simple transitive clauses. Greenberg's first word order universal was that the basic word order in languages is "almost always" one in which the subject precedes the object. That is, of the six logically possible orderings among V(erb), S(ubject), and O(bject), only three, VSO, SVO, and SOV,

---

[1] A third type of explanation can be found in the generative tradition, consisting of abstract representations and principles from which observable facts can be derived. I consider this sort of explanation incomplete without a grounding of the abstract representations and principles in human psychology, biology, or communicative needs. Paradigm examples of the generative approach applied to word order can be found in Chomsky (1970) and Kayne (1995).

occur with substantial frequency among the languages of the world. This has been confirmed in studies involving many more than Greenberg's original sample of 30 languages (see for example Hammarström 2016).

To explain why natural languages overwhelmingly favor SO over OS, it is necessary first to consider how linguists distinguish grammatical subjects from objects. Keenan's (1976) detailed examination of the properties of subjects contains many insights on this question. Two that are of particular relevance in the present context are the following:[2]

(1) "…subjects normally express the agent of the action, if there is one." (Keenan 1976: 321)

(2) "…subjects are normally the topic of the …sentence, i.e., they identify what the speaker is talking about." (Keenan 1976: 318)

Taken together with the many grammatical properties Keenan identifies as characteristic of subjects, what (1) says is that noun phrases denoting agents characteristically will exhibit those properties. In clauses denoting actions involving two participants (arguably the canonical type of transitive clause), the agent, as the perpetrator of the action, is psychologically more salient than the other participant. Greenberg's universal then has a fairly straightforward explanation: all else being equal, speakers will mention the more salient participant before the less salient one. Support for this comes from the studies described by Goldin-Meadow et al. (2008), in which participants given the task of describing an event using only gestures consistently produced gestures to denote agents before those used to denote patients.

Similarly, (2) helps to explain Greenberg's universal. Linguistic communication is not limited to isolated sentences. Rather, people usually speak in connected discourses, in which the connections between sentences can be as important to understanding as are the relations among the components of a sentence. It makes sense, therefore, to link an utterance to what preceded it early in each sentence. This promotes discourse coherence.

There is of course no guarantee that the semantic and discourse functions of subjects given in (1) and (2) will always converge. Hence, exceptions to both generalizations are not hard to find, and some marked constructions in lan-

---

[2] Keenan's discussion refers to "b-subjects," where the b is short for "basic". That is, he is concerned with subjects of basic, unmarked clauses (which he calls "b-sentences"), excluding passives and other marked constructions.

guages serve to indicate such mismatches. Both, however, help to explain why the basic word order in almost all languages has subjects preceding objects.

Among the three orders in which the subject precedes the object, languages with SOV as their basic order are the most common, followed by SVO languages, with VSO being considerably less common. Intriguingly, in the Goldin-Meadow et al. (2008) studies cited above, participants overwhelmingly produced gestures denoting the action after having produced gestures denoting the agent and patient. That is, they used the analogue to SOV ordering over 80% of the time, even if that was not the dominant order of their native languages. This suggests that the high proportion of languages that are SOV might be attributable to some deeper cognitive preference. Exploring this speculation further goes beyond the scope of this chapter.

Why are VSO languages less frequent than SOV or SVO languages? One possible explanation is based on a generalization due to Behaghel (1932: 4), "daß das geistig eng Zusammengehörige auch eng zusammengestellt wird."[3] Keenan (1979) argues that the denotation of the object of a transitive verb restricts the meaning of the verb in a way that the subject of a transitive verb typically does not. He cites (along with several other examples) the different types of actions denoted by *cut* when its object is a body part (*cut a finger*), a substance (*cut heroin*), or an abstraction (*cut prices*). This close semantic association between transitive verbs and their objects, combined with Behaghel's generalization above suggests that word orders in which the verb and object are adjacent should be preferred. Of the three basic word orders observing the preference for subjects before objects, only VSO separates the verb and the object.

Among Greenberg's other observations about cross-language word order regularities is that the position of the verb with respect to the object in the basic word order tends to correlate strongly with other ordering preferences. For example, VO languages tend to have prepositions, while OV languages tend to have postpositions; and genitives follow nouns in VO languages, but precede them in OV languages. Languages that conform to such tendencies are often referred to as head-initial or head-final languages, and consistency of head position across phrases is sometime called harmonic word order.

---

[3] In Wasow (2002) I translated this as, 'that what belongs close together conceptually also gets placed close together.' Behaghel labels this "[d]as oberste Gesetz" ('the highest law') of word order.

3 Language-internal order variation

Some of the factors influencing word ordering across languages also play a role in word order variation within languages. For example, in a normally SVO language like English, OSV sentences may be used to promote discourse coherence, as in the following:

(3) A: Do you know the Smiths?
     B: Him I've met; her I haven't.

Here B uses two OSV clauses in order to highlight the link between the objects of those clauses and the object in A's question, and to emphasize the contrast between the two parts of the answer. In this case, the violation of the general preference for subjects before objects in English serves a discourse function, much as the canonical ordering does most of the time.

Differences in word order within a language very often serve to signal some semantic or pragmatic difference. If B in example (3) responded with the same words, but using the normal SVO order, the meaning conveyed would be subtly different, focusing attention more on the speaker and less on the two other individuals mentioned. Bolinger (1968: 127) went so far as to claim "a difference in syntactic form always spells a difference in meaning." I believe this is an overstatement (see Wasow 2009: 262), but it is true that different word orders are very often used to convey different nuances of meaning.

3.1 Weight

Another important factor influencing word order variation is ease of processing. In three books (Hawkins 1994, 2004, 2014) and numerous papers, Hawkins posited several principles of efficient processing that he claims influence the grammatical forms speakers use. The one of most relevance to word order is what he calls Minimize Domains (MiD). Hawkins's formulation of MiD is fairly complex, involving supporting definitions, but the concept is quite intuitive: if there is some syntactic and/or semantic dependency between two elements, processing that dependency will be facilitated when they are in close proximity. In most sentences, there are multiple pairs of elements with such dependencies, and, according to MiD, word orders will be favored to the extent that they minimize the average distance between dependent pairs of elements.

Notice that this is an extension of Behaghel's generalization above. Behaghel's observation was evidently limited to semantic dependencies ("geistig eng Zusammengehörige"), but Hawkins extends this to syntactic dependencies, such as agreement and case marking, which do not always have a clear semantic basis.

One application of MiD is as an explanation and extension of another of Behaghel's generalizations. He dubbed this one, first published in Behaghel (1909/10), "das Gesetz der wachsenden Glieder" ('the law of the growing constituents'). Based on his examination of word order patterns in a number of European languages, he concluded (Behaghel 1932: 6) "daß von zwei Gliedern, soweit möglich, das kürzere vorausgeht, das längere nachsteht."[4] This generalization, sometimes known as "end weight" or "short-before-long" has been widely discussed in the literature. Hawkins (1994) sought to explain it in terms of processing, using a predecessor to (and special case of) MiD that he called Early Immediate Constituents (EIC). The idea was that processing efficiency would be maximized by minimizing the number of words that needed to be processed for the determination of immediate constituent structure.

As an illustration, in the two examples in (4), taken from Hawkins (2003), the underlined portion indicates the words that must be processed to determine that the verb phrase contains a verb and two prepositional phrases.

(4) a. The gamekeeper [VP looked [PP through his binoculars] [PP into the blue but slightly overcast sky]]
     b. The gamekeeper [VP looked [PP into the blue but slightly overcast sky] [PP through his binoculars]]

The underlined portion in (4a) is five words long, whereas in (4b) it is nine words long. Hence, according to Hawkins, structures like (4a) should occur more frequently and be processed more easily than structures like (4b). Notice that the VP in (4a) obeys Behaghel's *Gesetz der wachsenden Glieder*, whereas the VP in (4b) does not.

Hawkins (1994) deduced predictions from EIC and then checked corpus data from ten languages to test their accuracy. One novel prediction was that in consistently head-final languages, longer constituents should precede rather than follow shorter ones. To see why that is the case, consider the analogue to (4) in an SOV language with postpositions, rather than prepositions. The two VPs would look schematically like (5).

---

4 'that of two constituents, to the extent possible, the shorter one precedes, the longer one follows.'

(5) a. [vp [pp X X P] [pp X X X X X X P] V]]
    b. [vp [pp X X X X X X P] [pp X X P] V]]

Here the Xs stand for arbitrary words, the Ps are postpositions, and V is a verb. As in (4), the underlined regions represent the portion of the sentences that must be processed to determine the top-level constituent structure of the VP. But in this case, the underlined region is shorter in the (b) example. Thus, EIC predicts that consistently head-final languages should show a preference for long constituents before short ones, contrary to Behaghel's generalization. Two of the languages Hawkins (1994) applied EIC to were Japanese and Korean, which are consistently verb-final and postpositional; they did indeed show the long-before-short pattern. Subsequent corpus and psycholinguistic studies (e.g., Yamashita and Chang 2001) have supported this finding.

Gibson (1998, 2000) put forward a similar processing-based account of such phenomena (which I will henceforth refer to as 'weight effects'), grounding his account on properties of short-term memory. There are subtle but significant differences between Hawkins's and Gibson's proposals for how to compute dependency length, but they are conceptually very similar.

Faghiri and Samvelian (2020) argue that these differences lead to different predictions regarding the word order preferences in a verb-final but prepositional language (a combination that Greenberg (1963) noted was relatively uncommon). The reason for this can be seen by considering the two schematic structures in (6).

(6) a. [vp [pp P X] [pp P X X X X X] V]]
    b. [vp [pp P X X X X X] [pp P X] V]]

In cases like this, where recognition of the three immediate constituents of VP requires the full length of the VP, irrespective or order, Hawkins suggests that, at each word, the ratio of immediate constituents recognized to words processed should be computed, and then averaged over the whole structure. He calls this the IC-to-Word ratio. In (6), the calculations would be as follows:

(7) a. [vp [pp P X] [pp P   X   X   X   X X] V]]
           1/1 1/2   2/3 2/4 2/5 2/6 2/7 2/8 3/9   IC-to-Word ratio = 47.4%
    b. [vp [pp P   X   X   X   X   X] [pp P X] V]]
           1/1 1/2 1/3 1/4 1/5 1/6   2/7 2/8 3/9   IC-to-Word ratio = 36.8%

Hawkins's theory says that the higher IC-to-Word ratio (that is, (6a)) should be preferred. Gibson's account, on the other hand, simply compares the sum of the dependency distances in the two structures, with the lower total being preferred. In this case, that would be 8+6=14 in (6a) and 8+2=10 in (6b). Thus,

Hawkins predicts that a language with structures like those in (6) should prefer the short-before-long (a) order, whereas Gibson predicts it should prefer the long-before-short (b) order. Faghiri and Samvelian offer Persian as a language of this type and provide evidence that the long-before-short order is in fact preferred.

In the last decade, a number of linguists have applied computational tools to test the hypothesis that minimization of dependency length is an important factor in word ordering across languages (e.g., Gildea and Temperley 2010; Futrell et al. 2015, 2020). This research has taken advantage of the availability of corpora from an increasing number of languages that have been parsed for syntactic dependencies. In particular, the Universal Dependencies project (see https://universaldependencies.org/) has provided dependency-parsed corpora for dozens of languages. In each language, then, the dependency lengths in the corpus can be compared to the dependency lengths that would result if the ordering of the words in the sentences of the corpus is scrambled. Multiple such randomly scrambled alternatives are constructed for each corpus and compared to the original. In the languages that have been tested in this way, the total dependency lengths in the scrambled versions are consistently greater than in the original corpus, providing support for Hawkins's and Gibson's insight that dependency minimization is an important factor in determining word ordering. For a survey of a variety of types of research on dependency minimization, see Liu et al. (2017).

A fascinating next step in this research program is described by Hahn et al. (2020). They explored the hypothesis that some word order generalizations could be explained in terms of abstract properties of communication systems, independent of who or what is communicating. While earlier research advocating some version of dependency minimization as an explanation of word order facts cited human short-term memory limitations as the motivation for dependency minimization, Hahn et al.'s results do not rest on any assumptions about human psychology. Rather, they define a mathematically precise notion of communicative efficiency and argue that the actual word orders in the languages whose corpora they tested scored better on their metric of communicative efficiency than scrambled versions of the corpora. Communicative efficiency in their models correlated strongly (and inversely) with dependency length, suggesting that dependency minimization may be derivable from very general considerations about efficient communication. Moreover, in their models, languages with harmonic (that is, consistently head-initial or head-final) word orders scored higher on their metric of communicative efficiency than those whose basic word order is not harmonic.

(4) and (5) illustrate the effects of dependency minimization in the relative ordering of two PPs. Similar effects can be observed in the ordering of constituents of different types. Three alternations in English that have received considerable attention with regard to the effects of grammatical weight are so-called heavy NP shift (HNPS), the verb-particle construction, and the dative alternation. Examples are given in (8).

(8)   a.   I put the keys to my house and car on the table. ~ I put on the table the keys to my house and car.
       b.   I picked my car keys up. ~ I picked up my car keys.
       c.   I handed my car keys to the valet. ~ I handed the valet my car keys.

Each of these has idiosyncrasies, and Melnick (2017) has shown that they differ in their sensitivity to grammatical weight. In the case of HNPS, there is strong bias against the shifted ordering – that is, with the direct object non-adjacent to the verb. Even when the object is longer than the other postverbal constituent by as many as 3 or 4 words, corpora show a preference for the canonical ordering of V-NP-X (where X is often a PP, but may be of some other type). The canonical ordering is favored by the fact, discussed above, that verbs and their direct objects have a close semantic relationship. Hence, the shifted ordering is relatively rare. Indeed, in experimental studies in which participants are asked for acceptability judgments, the shifted order never gets higher average scores than the canonical order, no matter how large the length difference between the postverbal constituents (see Medeiros et al. 2020).

The verb-particle construction differs from most other weight-sensitive phenomena in that one of the two elements whose order can be reversed consists of a single (usually monosyllabic) word. Consequently, according to EIC, the ordering in which the particle is adjacent to the verb should be preferred in all cases except when the object NP is a single word. However, while corpus studies have shown a powerful weight effect, examples with the object NP intervening between the verb and the particle are by no means rare, provided that the object is under five words long (see, e.g., Lohse et al. 2004). In this construction, as in HNPS, the tendency for the object NP to appear adjacent to the verb even when considerations of weight would lead one to expect it to occur later in the verb phrase can plausibly be attributed to the close semantic connection between verbs and their objects. Lohse et al. pointed out that, in this construction, there is sometimes also a semantic dependency between the verb and its particle. To illustrate this, consider examples like *lift the glass up*, in which the glass is lifted and goes up, with *figure the problem out*, in which the problem cannot be said to be figured or to go out. Dependency minimization predicts that verb-particle pairs exhibiting such a dependency should appear adjacent (with the object following) at higher rates than verb-particle pairs without such a semantic dependency. The corpus study of Lohse et al. confirmed this prediction.

The dative alternation has two idiosyncrasies. The first is simply that the two variants differ not just in word order, but also in the presence or absence of a preposition. The second is that there are subtle meaning differences between the variants, as noted by Green (1974) and Oehrle (1976). For example, the contrast between (9) and (10) shows that the object of *to* in the prepositional variant may be either a recipient or a location, whereas the first object in the double object variant must be a recipient.

(9) a. The pitcher threw the ball to Jones.
   b. The pitcher threw Jones the ball.

(10) a. The pitcher threw the ball to first base.
   b. *The pitcher threw first base the ball.

In spite of these complicating features of the dative alternation, it exhibits clear weight effects in corpus data. This was noted by Hawkins (1994), and corroborated by Wasow (2002) and Bresnan et al. (2007), among many others. The latter paper provides an especially careful simultaneous examination of multiple factors that have been claimed in the literature to influence whether a speaker uses the prepositional or double object dative construction. It revealed not only that weight plays an important role in this alternation, but also that many other factors do too.

3.2 Information structure

One of these other factors is what is sometimes called "information structure" – that is, the distinction between information already available in the discourse and information that is newly introduced. There is a large literature on how word order is influenced by information structure, most of which claims a universal tendency to order old (also known as "given") information before new information.[5] This ordering seems natural in terms of discourse coherence, since the old information in a sentence provides a link to what was said earlier. Bres-

---

5 As I noted in Wasow (2002: 62–65), this literature can be quite confusing, owing to a plethora of terms used for information structural distinctions, uncertainty about whether different terminologies denote exactly the same distinction, and occasional contradictory claims about how information structure correlates with word order.

nan et al.'s (2007) multivariate models of the English dative alternation in the Switchboard corpus showed a strong effect of information structure, so that sentences with new theme arguments tended to occur in the double-object construction and those with new recipient arguments tended to occur in the prepositional construction.

A natural question that arises is whether the influence of information structure and grammatical weight are in fact distinct phenomena. Since old information can, in general, be comprehensibly expressed with fewer words than are required for new information, could it be that weight effects can be reduced to a preference for ordering old information before new? Arnold et al. (2000) investigated that question via a corpus study and an experiment, finding that the two influences on word order could in fact be teased apart. Bresnan et al.'s (2007) study reinforces this conclusion, since both information structure and weight emerge as significant factors in their model.

This finding should not have been surprising, for the two factors (weight and information structure) appear to differ in their cross-linguistic patterning. As noted above, the influence of weight on word order depends on the basic word order of the language: head-final languages prefer long-before-short ordering, whereas other types of languages seem to prefer short-before-long. But none of the literature on information structure claims any such bifurcation in the effects of information newness. The tendency to place old information before new information does not appear to depend on what the basic word order of the language is. And if weight effects are best explained in terms of processing while information structural effects are best explained in terms of discourse coherence, there is no reason to expect that their influence will always converge.

3.3 Accessibility

Another factor that has been found to influence word order is animacy. Corpus and experimental studies in a variety of languages have shown that there is tendency for expressions denoting animate entities to precede those denoting inanimates (see, e.g., Branigan et al. 2008). Indeed, Tomlin (1986: 102) put forward "The Animated First Principle," which he summarizes as: "In a transitive clause, all other things being equal, there is a tendency for the most 'animated' NP to precede other NPs."

This generalization overlaps to a significant extent with the observation, noted above, that agents tend to precede patients, for agents are typically (though not always) animate. Animate entities are psychologically highly salient. Hence, the tendency for expressions denoting animates to occur early seems natural from a production perspective. Spoken communication is highly time-constrained; if a person speaks too slowly or hesitantly, someone else will begin to speak. It makes sense, then, for speakers to utter what is readily accessible early, providing extra time to access and articulate what is less accessible. Branigan et al. (2008: 177), after reviewing a considerable body of literature on the subject, concluded that "animacy exerts its influence because it correlates with conceptual accessibility, an index of how easily a concept is retrieved from memory."

Notice that this idea also can be invoked as at least part of the explanation for the old-before-new preference, since old information is generally more accessible than new information. One might be tempted also to try to treat the short-before-long preference in many languages as another manifestation of accessibility, but this runs afoul of the fact that some languages exhibit a long-before-short preference. Yamashita and Chang (2001) investigated this issue and came to the conclusion that both conceptual accessibility and formal factors like length play a role in word order preferences.

4 Conclusion

Languages are primarily vehicles of spoken communication among humans. That fact alone goes a long way towards explaining many properties of language, including many of the regularities concerning word order that linguists have discovered. A variety of factors influence the sequencing of words in an utterance, including properties of the language being spoken, properties of the situations being referred to, the linguistic forms used in referring to those situations, what was said earlier in the discourse, and the state of mind of the speaker. There remain many puzzles about word order in particular languages and across languages, but great progress has been made in the last few decades both in documenting what word orders occur and explaining their distributions.

Note: Comments on earlier versions of this chapter by John A. Hawkins, the editors of this volume, and an anonymous reviewer led to various improvements. I am grateful for their help.

References

Arnold, Jennifer, Anthony Losongco, Thomas Wasow & Ryan Ginstrom. 2000. Heaviness vs. newness: The effects of structural complexity and discourse status on constituent ordering. *Language* 76(1). 28–55.

Behaghel, Otto. 1909/10. Beziehungen zwischen Umfang und Reihenfolge von Satzgliedern. *Indogermanische Forschungen* 25. 110–142.

Behaghel, Otto. 1932. *Deutsche Syntax: Eine geschichtliche Darstellung. Band IV: Wortstellung. Periodenbau.* Heidelberg. Carl Winters Universitätsbuchhandlung.

Bolinger, Dwight. 1968. Entailment and the meaning of structures. *Glossa* 2. 119–127.

Branigan Holly P., Martin J. Pickering & Mikihiro Tanaka. 2008. Contributions of animacy to grammatical function assignment and word order during production. *Lingua* 118. 172–189.

Bresnan, Joan, Anna Cueni, Tatiana Nikitina & Harald Baayen. 2007. Predicting the dative alternation. In Gerlof Bouma, Irene Kramer & Joost Zwarts (eds.), *Cognitive foundations of interpretation*, 69–94. Amsterdam: Royal Netherlands Academy of Science.

Chomsky, Noam. 1970. Remarks on nominalization. In Roderick A. Jacobs & Peter S. Rosenbaum (eds.), *Readings in English transformational grammar*, 184–221. Waltham, MA. Ginn & Co.

Faghiri, Pegah & Pollet Samvelian. 2020. Word order preferences and the effect of phrasal length in SOV languages: Evidence from sentence production in Persian. *Glossa* 5(1). 86.

Futrell, Richard, Kyle Mahowald & Edward Gibson. 2015. Large-scale evidence of dependency length minimization in 37 languages. *Proceedings of the National Academy of Sciences* 112(33).

Futrell, Richard, Roger Levy & Edward Gibson. 2020. Dependency locality as an explanatory principle for word order. *Language* 96(2). 71–412.

Gibson, Edward. 1998. Linguistic complexity: Locality of syntactic dependencies. Cognition 68. 1–76.

Gibson, Edward. 2000. The dependency locality theory: A distance-based theory of linguistic complexity. In Alec Marantz, Yasushi Miyashita & Wayne O'Neil (eds.), *Image, language, brain: Papers from the first mind articulation project symposium*, 94–126. Cambridge, MA: MIT Press.

Gildea, Daniel & David Temperley. 2010. Do grammars minimize dependency length? *Cognitive Science* 34(2). 286–310.

Goldin-Meadow, Susan, Wing Chee So, Asli Özyürek & Carolyn Mylander. 2008. The natural order of events: How speakers of different languages represent events nonverbally. *Proceedings of the National Academy of Sciences* 105. 9163–9168.

Green, Georgia. 1974. *Semantics and syntactic regularity*. Bloomington: Indiana University Press.

Greenberg, Joseph. 1963. Some universals of grammar with particular reference to the order of meaningful elements. In Joseph Greenberg (ed.), *Universals of language*, 3–113. Cambridge, MA: MIT Press.

Hahn, Michael, Dan Jurafsky & Richard Futrell. 2020. Universals of word order reflect optimization of grammars for efficient communication. *Proceedings of the National Academy of Sciences* 117(5).

Hammarström, Harald. 2016. Linguistic diversity and language evolution. *Journal of Language Evolution* 1(1). 19–29.

Hawkins John A. 1994. *A performance theory of order and constituency*. Cambridge, UK: Cambridge University Press.
Hawkins, John A. 2003. Efficiency and complexity in grammars: Three general principles. In John Moore & Maria Polinsky (eds.), *The nature of explanation in linguistic theory*, 121–152. Stanford: CSLI Publications
Hawkins, John A. 2004. *Efficiency and complexity in grammars*. Oxford: Oxford University Press.
Hawkins, John A. 2014. *Cross-linguistic variation and efficiency*. Oxford: Oxford University Press.
Kayne, Richard. 1995. *The antisymmetry of syntax*. Cambridge, MA: MIT Press.
Keenan, Edward. 1976. Towards a universal definition of "subject". In Charles Li (ed.), *Subject and topic*, 303–333. Cambridge, MA: Academic Press.
Keenan, Edward. 1979. On surface form and logical form. *Studies in the Linguistic Sciences* 8(2). 163–203.
Liu, Haitao, Xu Chunshan & Liang Junying. 2017. Dependency distance: A new perspective on syntactic patterns in natural languages. *Physics of Life Reviews* 21. 171–193.
Lohse, Barbara, John Hawkins & Thomas Wasow. 2004. Domain minimization in English verb-particle constructions. *Language* 80(2). 238–261.
Medeiros, David J., Paul Mains & Kevin B. McGowan. 2020. Ceiling effects on weight in heavy NP shift. Accepted for publication in *Linguistic Inquiry*, available online at https://doi.org/10.1162/ling_a_00382
Melnick, Robin. 2017. *Consistency in variation: On the provenance of end-weight*. Stanford University Ph.D. thesis. https://www.academia.edu/39400150/Consistency_in_variation_On_the_provenance_of_end-weight
Oehrle, Richard T. 1976. *The grammatical status of the English dative alternation*. MIT Ph.D. thesis. https://pure.mpg.de/rest/items/item_405081/component/file_405080/content
Simpson, Jane. 2007. Expressing pragmatic constraints on word order in Warlpiri. In Annie Zaenen, Jane Simpson, Christopher Manning & Jane.. Grimshaw (eds.), *Architectures, rules, and preferences: A festschrift for Joan Bresnan*, 403–427. Stanford: CSLI Publications.
Song, Jae Jung. 2012. *Word order*. Cambridge, UK: Cambridge University Press.
Tomlin, Russell. 1986. *Basic word order*. London: Routledge.
Wasow, Thomas. 2002. *Postverbal behavior*. Stanford: CSLI Publications.
Wasow, Thomas. 2009. Gradient data and gradient grammars. *Proceedings of the 43rd Annual Meeting of the Chicago Linguistics Society*, 255–271.
Yamashita, Hiroko & Franklin Chang. 2001. 'Long before short' preference in the production of a head-final language. *Cognition* 81(2). B45–B55.

Doris L. Payne
Proximal to distal: Information flow and order in Maa

Abstract: A long-standing issue in discourse-pragmatics is the differing conceptual content linguists have for information-structural terms related to "topic" and "focus". This paper aims to clarify some concepts relevant to this domain, including (basically) deixis terms like "proximal" (or "proximate") and "distal", which may be relevant to the linear placement of what this volume calls Target phrases (approximately, RECIPIENT, GOAL, BENEFACTIVE arguments). The paper shows how selected notions of *topic* might be operationalized to investigate relative order of object phrases encoding these and INSTRUMENT phrases in Maa (Eastern Nilotic) ditransitive clauses. On the basis of a corpus study, I suggest that relative to postverbal phrases, Maa has a general constructional pattern of ordering cognitively proximal information before cognitively distal information.

Keywords: deixis; ditransitive; Maasai; Nilotic; topic

1 Background and goals

What determines linear order of phrases can vary depending on the language and construction. Order may correlate with grammatical roles like subject versus object, animacy, phonological weight, syntactic complexity, information structure or specific discourse-pragmatic function, or features of the historical origins of a construction. If order in a (set of) construction(s) demonstrates a strong relationship with particular discourse-pragmatic statuses, the relationship should be considered part of the speaker's routinized grammar (Payne 1992, 1995, 1999), certainly within a construction grammar view of syntax. Traditionally, discourse-pragmatics concerns the speaker's cognitive context, her assumptions about the hearer's cognitive context, communicative intentions and actions, presupposed versus asserted information, deictic issues, and information flow. The distinctness of pragmatics from semantics is debated (Stalnaker 1999; Korta and Perry 2020), and I do not make a strong distinction here.

Doris L. Payne, Department of Linguistics, University of Oregon, Eugene, USA,
E-mail: dlpayne@uoregon.edu

"Targets", as the term is used in this volume, are arguments that roughly correspond to RECIPIENTS, GOALS, ADDRESSEES, and BENEFICIARIES. In Maa (Eastern Nilotic), these as well as INSTRUMENTS are usually expressed as core-argument objects, shown by their marking as objects in the verb and by word order. In part, this paper explores what might account for relative order of objects in Maa ditransitive clauses. I refer to any non-subject core argument as an object (O), regardless of semantic role, and use the term R (reminiscent of RECIPIENT) in place of "Target".

Maa designates a number of Eastern Nilotic linguistic varieties spoken in Kenya and Tanzania. This study specifically draws on data from Maasai and Ilchamus Maa varieties. Since the data come from more than one region, I use the term *Maa* throughout the paper.[1]

This paper first takes the opportunity to clarify selected senses of the terms *focus, topic*, and other closely-related notions including some in the realm of deixis. This is done in Section 2. Section 3 introduces order patterns in four Maa ditransitive clause types. Section 4 quantitatively explores selected factors that may bear on order of the postverbal Os. Section 5 summarizes the main results and compares them to what might motivate VSO versus VOS order in monotransitive clauses.

2 Concepts and methodology

An endemic issue in discourse-pragmatics is the differing definitions and methodologies (or no clear methodology) for identifying information structure statuses. I comment on selected notions that have been associated with the terms *focus, topic*, and *proximal* (or *proximate*) versus *distal*, which are relevant to the domains of attention and information flow. Here I am concerned with cognitive or informational status, and not primarily with morphosyntactic notions.[2]

---

[1] ISO 639-3 codes for Maa varieties are saq, mas, and nsg (Lewis et al. 2014); "mas" encompasses Maasai, Arusa, and Parakuyo varieties. To my knowledge, the issues discussed here do not vary across varieties. The Maa research has been partially supported by NSF grant SBR-9616482 and Fulbright awards for work in Kenya and Tanzania. I am grateful to many colleagues and friends, especially Leonard Ole-Kotikash, A. Keswe Mapena, and Sarah Tukuoo (https://pages.uoregon.edu/maasai/). An earlier version of some of these results was published in Payne (2015). I am grateful to Osamu Hieda for spurring aspects of this research.

[2] *Proximate/obviative* and *direction* (*systems*) sometimes refer to morphosyntactic forms that at least partially correspond to proximal/distal conceptual or cognitive statuses, and which can relate to issues of discourse topicality (Dryer 1992; Zúñiga 2006: 29–46).

First, let us consider "focus". Some italicized expressions in (1) involve the word *focus*, and others are included in discussions of "focus", but they do not all designate the same cognitive phenomenon (see also Matić and Wedgwood 2013), though some overlap (e.g. the "evocation of alternatives" may underlie "contrastive focus").

(1) **Selected senses of the term *focus***
  a. Non-contrastive *focus of assertion* (Lambrecht 1994); roughly similar to *rheme* (Firbas 1964; cf. Vallduví and Vilkuna 1998).
  b. Contrastive or marked focus subtypes: *exclusive focus, restricting focus, expanding focus, predicate-centered focus, argument focus, polar focus,* counter-expectation, etc. (Watters 1979; Dik et al. 1981; Vallejos 2009).
  c. The evocation of *alternatives*, often with selection of an alternative (Rooth 1992; Vallduví andVilkuna 1998).
  d. *Cognitively salient* or vivid (human, animate, large size, imageability, standing out against a ground, etc.) (Myachykov 2007: 29, 39).
  e. Bringing something into the cognitive laser-like or "spotlight" *focus of attention;* attentional shift (Posner and DiGirolamo 1998; Tomlin 1995; Myachykov 2007: 23).
  f. Being in the cognitive "spotlight" *focus of attention* (Posner and DiGirolamo 1998; Tomlin 1995; Myachykov 2007: 22). This may be related to *activated* status in the information-structure literature (Chafe 1976), and concerns information currently being processed or held in short-term memory.

Information-structure claims in the domain of "topic" frequently fail to distinguish among the concepts in (2) and (3) (among possibly others; see also Jacobs 2001).

(2) **Selected senses of the term *topic***
  a. *(Discourse) topic:* A summarizing macro-proposition for a (section of) discourse (van Dijk 1977).
  b. *(Participant) topic:* A participant or objectified non-physical concept that a (section of) discourse is about (van Dijk 1977).
  c. *(Sentence) topic*: The participant or objectified non-physical concept that a sentence is about (Hockett 1958: 201; Reinhart 1982; see also van Dijk 1977).

The following are closely related to, if not used as identifying features of one or another sense of *topic*. Note, however, that some here (e.g. 3c) may overlap with some uses of the term *focus* (cf. 1f above). This reflects the long-standing difficulties in the linguistic literature about how these terms are used – e.g. using the same or similar terms for what may be distinct cognitive statuses.

(3) **Notions closely related to the term *topic***
   a. The *cognitive foundation* to which incoming information is related (Gernsbacher 1995; cf. Jacobs' (2001) *addressation* parameter).
   b. *Frame:* The participant or concept that delimits a context within which a proposition holds (Chafe 1976; cf. Fauconnier's (1994) *mental space*).
   c. Highly *accessible* or *activated* information that is currently being processed in short-term memory (cf. Myachykov 2007: 43–44). Chafe (1976) essentially equates this with *given/old* information. Chafe (1987) proposes that information can also be *semi-activated*.
   d. *Familiar* or *evoked* information (Prince 1981; Gundel et al. 1993). When information is familiar or evoked because it is already active in short-term memory, this would be equivalent to (3c). Familiar information may be part of the *common ground* (Stalnaker 1999), or what is presupposed.
   e. The participant higher in anaphoric or cataphoric *continuity* (Givón 1983).

I will not expand on all the notions in (1) through (3), but make a few comments that will be relevant to the Maa case study presented below.

Rather in line with van Dijk (1977), I reject the idea that one can, in any meaningful way, identify an information-structure status of "topic" on a strictly sentence- or proposition-internal basis. I do not find that sense (2c) is non-circularly operationalizable, except in relation to (2a–b). Brief consideration of some authentic English examples may help address why.

First, consider the sentence *A couple plans a grand opening for a brewpub*. This is the first sentence of a newspaper article, orthographically written as a subtitle.[3] It is unclear that this sentence has an internal phrase or concept that the rest of the sentence is "about"; upon processing this sentence alone, how can we say it is about 'a couple', 'a grand opening', 'a brewpub', 'plans for a grand opening', or even the entire proposition? The immediately-preceding main title is *Hard Knocks Life in Cottage Grove*, which, in my understanding even as a reader within the newspaper's distribution area, does not establish anything within *A couple plans a grand opening for a brewpub* as presupposed by the time one processes the subtitle. This is likely what Lambrecht (1994) calls

---

3 http://registerguard.com/rg/business/32502996-63/cottage-grove-brewpub-plans-grand-opening.html.csp, consulted December 3, 2014.

a thetic sentence, containing entirely asserted information and no presupposition. We thus may set aside thetic sentences as simply irrelevant to issues of "sentence topic".

But a second issue is presented by the next sentence of the newspaper article (the first in the article body): *In the next few weeks, Hard Knocks Brewing – which touts itself as "Cottage Grove's first and only brewpub" – will hold its grand opening on East Main Street, owners Ben and Kate Price say.* From a discourse-topic perspective (2a), this first body sentence is about something propositional, like 'Hard Knocks Brewing is planning a grand opening'. But what grounds are there for saying which element is "the sentence topic" (2c), about which the rest of the proposition makes a comment or assertion? If one were to claim that the sentence topic was 'Hard Knocks Brewing', that claim would almost certainly be biased by what is coded as the subject; just from a sentence internal perspective, one might equally suggest that the sentence topic is 'a grand opening'. But after examining the entire article, one could strongly argue that the participant this sentence is primarily about is either 'Ben and Kate Price' or 'Ben and Kate Price's business development'. Methodologically, we might closely examine propositional sequences or use experimental protocols to determine an "aboutness" relationship between a participant named in part of a sentence and the larger discourse or mental domain; but this is a discourse-based determination of "topic" in the (2b) sense – not a sentence-level determination. My point is that it is usually impossible to say that a sentence is "about" one participant or phrase in a sentence simply by examining that one sentence.

A *framing* element (3b) delimits a mental space (Fauconnier 1994) or cognitive world within which something else is relevant. A frame may be expressed in an NP, PP, or a (non-main) clause. In the sentence **With stay-at-home orders expiring and businesses reopening**, *all the scientific data is being scrutinized anew*,[4] the highlighted prepositional phrase delimits or roughly restricts the comprehender's interpretation of what scientific data is newly scrutinized. Due to the phrase *stay-at-home orders* in the 2020 news article from which this sentence comes, the reader does not likely consider that data gathered by, for instance, the Hubble telescope is being scrutinized anew, but rather data about the SARS-CoV-2 virus.

The notions in (3c–d) concern what is often called given/old information. Chafe properly grounds these in a memory and information-processing frame-

---

4 https://www.washingtonpost.com/health/tell-me-what-to-do-please-even-experts-struggle-with-coronavirus-unknowns/2020/05/25/e11f9870-9d08-11ea-ad09-8da7ec214672_story.html, consulted May 26, 2020.

work. Though cognitively activated (or given/old) or familiar concepts often designate information that is part of a presupposition, they are logically distinct from it. *Presupposed* means the speaker presents information in a manner such that the hearer is supposed to "take it for granted" without challenge, as part of the shared common ground (Lambrecht 1994; Stalnaker 1999, 2002). In the following from the *Hard Life Knocks* article, the proposition 'she is a Cottage Grove City Council member' is presupposed, despite being completely new: *In addition to being a Cottage Grove City Council member, she already owned and operated a fitness business in Cottage Grove.*[5]

Finally, (4) lists some factors underlying cognitively *proximal* versus *distal* statuses. Many deictic morphemes (demonstratives and pronoun forms, temporal adverbs, 'ventive', 'itive', etc.) may indicate something about literal space, time, or person, but they often metaphorically extend to cognitive closeness/farness, and bear on issues of attention and attention flow (DeLancey 1981; Zúñiga 2006: 30–32). For example, taking one participant's point of view (4f) is a deictic stance as that participant is conceptualized as cognitively "closer" to the speaker, metaphorically speaking, than another at some discourse moment. Cognitively *proximal* and *distal* statuses belong to a hybrid pragmatic-semantic domain, given the factors in (4).

(4) **Selected factors for cognitively *proximal/distal* statuses**
   a. physical or cognitive proximity/remoteness from a reference point
   b. direction 'towards'/'away from' a reference point
   c. animate/inanimate
   d. human/non-human
   e. speech-act (1st and 2nd)/non-speech-act (3rd) participants
   f. which participant's point of view is taken
   g. starting/ending point of attention flow
   h. current central interest/ancillary or less central
   i. high/low discourse saliency
   j. high/low topicality

The features in (4), perhaps especially (4g–j), connect to information structure (understood broadly) and to linear order. Relative to (4g), for instance, Zúñiga (2006: 30) states, "linguistic attention flow roughly corresponds to information structure in the sense of linear ordering of [elements] in a clause." For example,

---

[5] Negation can probe what is presupposed versus asserted. *No, she didn't* negates 'she already owned and operated a fitness business in Cottage Grove'. *No, she isn't*, for negating 'she is a Cottage Grove City Council member', is a pragmatically infelicitous rejoinder.

SOURCE may be iconically placed before GOAL; giver (AGENT) before THEME, before RECIPIENT. Attention flow and sometimes linguistic order advance from the current center of interest (4h) to something else.

Methodologies for exploring what statuses in (1) through (4) a given phrase has, and potential correlation of that status to grammatical patterns and devices, include qualitative and quantitative study of already-produced texts and experimental manipulation of production and comprehension (cf. Arnold et al. 2000 on English ditransitives).

With this general background, we turn to what might account for order of R and core INSTRUMENT phrases relative to other object (O) phrases in Maa ditransitive clauses. After examining various factors, I will propose that the information structure principle in (5) accounts for the vast majority of post-verb order facts (setting aside some potential focus factors).

(5) **Postverbal Order Hypothesis:** Proximal before Distal

3 Order of R and INSTRUMENT versus other Os in Maa ditransitive clauses

Maa is said to have verb-subject-object (VSO) basic order (Tucker and Mpaayei 1955: 7), though SVO, OVS, and VOS also occur. In ditransitives, all possible orders of THEME, R, and INSTRUMENT occur. There are conflicting claims about what motivates order variations. Using a topic-continuity methodology (Givón 1983), Payne et al. (1994) argue that VOS is a "pragmatic inverse" and observe that in the Hollis' (1905) texts used for their study, VOS always occurs if the O is pronominal. They interpret these facts as demonstrating a correlation between immediately postverbal position and more topical status. Koopman (2005: 291) claims that VOS occurs when the O is focused or prominent. Allan (1990) and Payne (1995) describe preverbal subject and O NPs as focused.

Before further examining Maa word order in ditransitives, a few comments are in order about Maa morphosyntax. Subjects and up to one singular speech-act-participant O can be marked by pronominal verb prefixes. Third person activated and/or identifiable Os are usually expressed by zero anaphors, and are not marked by verb prefixes. It is infrequent to find two, much less three core lexical NPs in a clause. Postverbal NPs contrast tonally for nominative versus absolute case (the latter is used for citation forms, accusative role, and other functions; there are multiple tone classes for case; Tucker and Mpaayei 1955).

Ditransitives allow two overt postverbal absolute-case Os. The Os are syntactically symmetrical, with no difference in relativization or other inter-clausal pivot properties. From the perspective of semantic (macro-)roles, order variation is possible and either O (but at most one) can be registered by a verb prefix.

Sections 3.1 and 3.2 illustrate various orders of Os, with brief comments on semantic-discourse features. Semantic macro-roles of Os are labeled, using the following terminology:

- THEME: Argument of a 3-argument predicate which literally or metaphorically moves; situation enabled or allowed.
- R: Argument of a 3-argument predicate that literally or metaphorically receives the THEME. R includes RECIPIENT, BENEFACTIVE, GOAL, ADDRESSEE, ENABLEE/ALLOWEE.

Two terms refer to applied Os. These are identified by specific applicative suffixes:
- INST(rument) = Applied O with the applicative *-ie(k)* (includes semantic INSTRUMENT, MEANS, CAUSEE; some LOCATIONS, TIME).
- DAT(ive) = Applied O with the applicative *-akɪ(n)* (allomorphs *-oki(n)*, *-ɪkɪ(n)*) (includes RECIPIENT, BENEFACTIVE, MALEFACTIVE, GOAL, item closely followed). The morphosyntactically-identified DAT(IVE) is a semantic R.

3.1 R/INST-THEME order

Examples (6)–(7) show R–THEME order with the ditransitive verb *ɪshɔ* 'give/allow'. In (6), R is human, referential and discourse-established; THEME is non-referential.[6]

(6)   VERB   R   THEME
n-éítʊ   *e-lotú*   *a-ɪshɔ́*   *m=kéra*   *ɛn=dáa*
CN1-NEG.PF   3-come.SG   INF.SG-give   FPL=children   FSG=food
'She did not come to give the children food.' (negligent mother, divorce.006)

In (7), R and THEME are non-referential, though R is human. The verb is in an impersonal non-promotional form (Payne 2011), and both postverbal NPs have

---

6 Free translations are followed by a phrase describing the discourse context, plus a text-line reference. High tone is marked by an accute accent, a raised down arrow marks downstep High, a circumflex marks Falling tone over a word-final mora (two orthographic circumflexes mark a diphthong), and Low tone is unmarked.

O properties: absolute case form, no preposition, and either (but only one at a time) could be marked in a verb prefix if it were first or second person singular.

(7)   VERB   R   THEME
     káke   mm-ɛ-ɪshɔr-í   il=ayîôk   ɔ=sárgɛ́
     but   NEG-3-give-IMPS   MPL=boys   MSG=blood
     'but (uncircumcised) boys are not given blood (to drink)' (distribution of slaughtered ox portions, inkiri.012)

In (8) with 'give' and (9) with 'put', R is a pronoun 'us' – a quintessentially familiar participant. In (8), R and THEME are human and/or divine, referential, and discourse-established, while THEME is identifiable and possessed.

(8)   VERB   R   THEME
     ...   n-í-nchɔɔ   iyíóók   ɛn=kéráí   inó   Yésu
             CN1-2-give   us   FSG=child   your   Jesus
     '… you give us your child Jesus.' (prayer, camus1.067b)

In (9), THEME is the abstract concept 'name' and is not a previously-established concept in the discourse.

(9)   VERB   R   THEME
     ɛ-nyaákɪtâ   áa-pɪk   iyíóók   nk=áí   árná
     3-do_again.PF.PL   INF.PL-put   us   FSG=other   name
     'they [the Tugen] have given us another name' (ethnic group interaction, camus4.300)

Example (10) has R–THEME order, with the dative applicative -oki(n) plus impersonal morphology on 'cut'. The dative creates core objects with "Target" roles such as BENEFACTIVE/MALEFACTIVE, GOAL-reached, etc.

(10)   VERB   R   THEME
     n-é-duŋ-**okin**-í   síî   [ɪl=tóŋáná lɛnyê]   [e=síáná   oó
     CN1-3-cut-DAT-IMPS   indeed   MPL=people his   FSG=number of.PL
     in=kíshú   naá-ítayu   naá-lak-ie   ol=dísí]
     FPL=cattle   REL.F-take_out   REL.F-pay-INST   MSG=DC
     'his people were sentenced ['cut for'] the number of cows that they take to pay with for the DC [district commissioner].' (British–Maasai conflict, DC.029)

Example (11a) illustrates INST–THEME order in a ditransitive formed with the instrumental applicative -ie(k). (We will discuss 11b later.)

(11)
a.

VERB		AGENT	
*m-e-ok-**ié***	*oshî*	[*ol=porrór*	*láŋ*]
NEG-3-drink-INST	always	MSG=age_set.NOM	our.NOM
INST	THEME		
in=kikompení	*sháai*		
FPL=cups	tea		

'Our age-set does not always drink tea using cups;'

b.

VERB	R	INST
*én-chor-**ie***	*iyíóók*	*il=oríkan*
PL.IMP-give-INST	us	MPL=seats

'give it [tea] to us on/using seats.' (enkang-enkai 0008b)

3.2 THEME–R/INST order

We now illustrate THEME-first ditransitives. In (12) with 'give', an inanimate THEME precedes an animate R.

(12)

VERB	THEME		
n-é-íshɔr-ɪ	*en=kelehé*	*aké*	*dúóó*
CN1-3-give-IMPS	FSG=meat.for.guests	just	relevant
R			
[*ɪl=tʊ́ŋánák*	*oó-po-ito*	]	
MPL=people	REL.MPL-go.PL-PROG		

'People just passing by are given meat-for-guests' (meat cuts, inkiri.068)

Examples (13)–(15) demonstrate THEME–R order with 'put'. In (13), THEME and R are non-referential, but THEME is human. It might be argued that R is a LOCATIVE, but this is just an issue of animacy of the R (compare 9 above).

(13)

VERB	THEME	R
m-á-tɪ-pɪk	*in=tóyie*	*sukúul*
OPT-1PL.SBJV-SBJV-put	FPL=girls	school

'Let's send girls to school' (old and new child-rearing traditions, aibart-isho.026c)

In (14), THEME and R are referential and previously mentioned in the text; but the THEME is animate and is a key character in the story.

(14)

VERB	THEME	R
n-é-pɪk-í	*ɔl=kítéŋ*	*ɔl=mʊnánda*
CN1-3-put-IMPS	MSG=bovine	MSG=marketplace

'The bullock was put into the marketplace.' (British and Maasai conflict, DC.021)

In (15b) the free-pronoun O *nɪnyé*, for an animate R, follows the lexical O *m=péré*; this order is highly unusual for pronouns. The context is somewhat contrastive: normally, one would spear an animal. The situation is being set up so a warrior can be speared with plausible deniability about intention to spear him.

(15) a. néākō   nchɔ́ɔ   peê   e-lotú   ŋûês
         THEREFORE   allow   so   3-come.SG   beast.NOM
         a-ɨ́m   karíbu   nɪnyé
         INF.SG-pass welcome   3SG
         'Make it [lit. 'allow'] so that when a wild animal comes to pass by him [the warrior]'

   b.   VERB   THEME   R
         n-í-royíé   a-pík   m=péré   nɪnyé
         CN1-2-fight INF.SG-put FSG=spear 3SG
         'you spear him' [instead of the animal, and make it look like an accident] (conversations, camus4.360-361)

In (16), THEME–R order occurs with the dative applicative on transitive 'turn'. This contrasts with (10) above: in (10) the heavier NP occurs farthest from the verb, but in (16) the lighter NP is farthest out.[7] Consistent across both is that the animate participant is closest to the verb.

(16)   VERB   THEME
       áa-bam-ɪkɪ   [il=kúmok   tɛ   n=kɔ̂p   ɔ́ɔ   Lchámús]
       INF.PL-turn-DAT   MPL=many   OBL   FSG=land.NOM of.PL Chamus
       R
       [o=róréí   lɛ́   nkáí]
       MSG=word   M.of   God
       'to direct many in the land of Chamus to the word of God' (conversation, camus1.075)

In (17) and (18), the instrumental applicative occurs on transitive roots, and the INSTRUMENT O occurs after the THEME – unlike in (11a).

---

[7] The quantitative study in Section 4 does not systematically analyze the effect of heaviness on order, though it does evaluate pronominal vs. lexical status which may interact with heaviness. Examples like (15) and (16) would be counter-examples to an order principle like "light before heavy".

(17)

			VERB	THEME
m-ɛ-áta	ɔl=ayíóni	obô	lé-m-é-íbúŋ-ie	ményé
NEG-3-exist INST	MSG=boy	one.M	REL.M-NEG-3-touch-INST	father.PSD3SG

ɛ=ílátá
FSG=oil

'there is not one son that will not smear his father with oil.' (death customs, enkeeya1.020b)

(18)

VERB	THEME	INST	
tó-osh-ie	iyíóók	[ɛldê	muátɛ]
IMP.SG-hit-INST	us	that.M	deserted.home

'hit us [a cow with eight tails] against [lit. 'with'] that deserted home' [so you can take one tail away with you] (mythical, enamuke2.0020)

Finally, (11b) above is a four-argument clause with overt R as the first O, INSTRUMENT as the second O, and zero anaphora for the THEME 'tea'.

With this introduction to ditransitives, we now examine factors that might correlate with relative order of the two Os.[8]

4 Corpus study

The corpus study is based on 36 texts gathered between 1994 and 2014 (including personal and third-person narratives, expository, descriptive, and conversational discourse). I consider only clauses with the lexical predicates *ɪshɔ(r)* 'give/allow' and *pɪk* 'put'; or monotransitive roots with dative *-akɪ(n)* or instrumental *-ie(k)* applicatives which create ditransitive stems.[9] The data was sifted to identify ditransitives with two overt postverbal O phrases. This yielded 67 clauses: 31 'give/allow', 10 'put', 16 dative, and 10 instrumental.

---

8 Four-argument clauses are so infrequent that relative order of three O's cannot be meaningfully explored in the corpus.
9 Constructions with these stems plus middle or antipassive morphology are excluded, as these reduce the number of Os. The impersonal construction does not reduce the number of Os, so instances of this construction are included.

4.1 Order of semantic (macro-)roles

Figure 1 evaluates whether there is a preference for R to be the first O and THEME to be the second O. For each ditransitive construction type, Figure 1 shows the percentage of V-R/X-THEME/X clauses; X is any role other than R or THEME (i.e. V-R-INST, V-R-THEME, and V-INST-THEME orders). For instance, 97% (30 out of 31) 'give' examples had V-R-THEME order, and one had V-THEME-R order. In contrast, 30% of 'put' examples had V-R-THEME order.

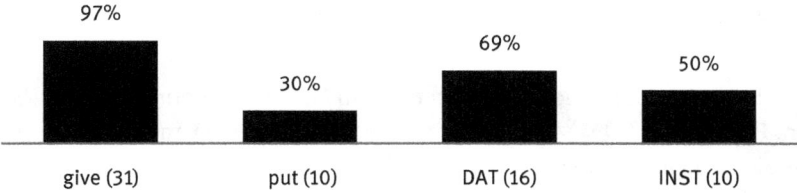

Figure 1: Ditransitive clauses with V-R/X-THEME/X order.

Figure 1 shows three things: 1) Order of macro-roles varies in all four ditransitive types. 2) Order variation is more prominent (i.e. does not tend toward one extreme or the other) in the applicative ditransitives than in the lexical 'give/allow' and 'put' ditransitives. 3) 'Give' and 'put' tend toward opposite R and THEME orderings. We comment more on these results in Section 5.

4.2 Free pronominal versus lexical status

Figure 1 (and Section 3) show that semantic (macro-)role does not provide a 100%-generalization about what governs order between the Os in a ditransitive clause. This section evaluates order of ditransitive Os relative to free pronoun vs. lexical status. Fifteen of the 67 ditransitives had a free pronoun O. Figure 2 displays these by type (number, person, reflexive) in NP1 and NP2 position. NP1 is the first O after the verb, and NP2 is the subsequent O.

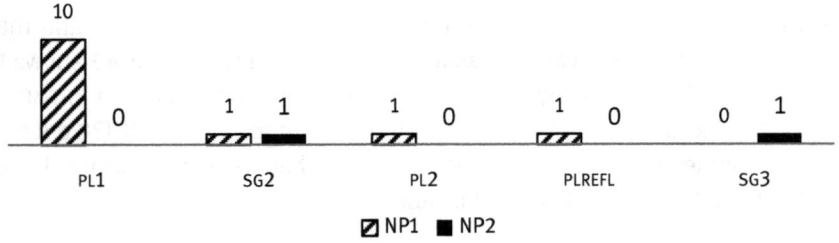

Figure 2: Free O pronouns in ditransitive clauses (N = 15).

Figure 2 shows a strong tendency for free ditransitive O pronouns to be in NP1 position. Payne et al. (1994) found that in monotransitive VSO and VOS clauses, this was also true for pronous in subject versus O roles.

1SG and 2SG are rarely expressed as free pronouns in the language unless there is some marked focus. This is because pronominal verb prefixes register singular speech-act-participants. But marked focus is not necessary for a free pronoun, as seen in (8) above and (19). Since activated and identifiable 3rd person Os are typically null, they are rare as free pronouns outside of marked focus.

(19)   *ɛnk=áí*   *papâ*   *mɪ-kí-naur-u*
       FSG=God   father   NEG-1PL-be.tired-INCHO

	VERB	R	THEME		
kɪ-gɪrá	*áa-ɪshɔ*	*iyíé*	*ɛn=ashê*	*sápʉk*	*olêŋ*
1PL.WHILE-PROG	INF.PL-give	2SG	FSG=thanks	big	very

'God (our) father, we will not tire while we are giving you a lot of thanks' (prayer, Camus2.150)

Free O pronouns can nevertheless occur as NP2. The two instances of this in the corpus (see the black bars in Figure 2) are in marked contexts (see 15b above).

Verb prefixes do not mark 3rd person, 1PL and 2PL Os, nor 2SG O when the subject is 1PL (Payne et al. 1994). Thus, these Os are not distinguished unless a free object pronoun is used. This is perhaps why *iyíé* occurs in (19). The high number of 1PL free pronoun Os (N=10) in Figure 2 is likely because otherwise it would be ambiguous whether the O was 1PL, 2PL (sometimes 2SG), or a 3rd person available in the context.

4.3 Animacy and topicality

We now examine order of ditransitive Os relative to animacy and participant topicality (2b). Referentiality was also considered, but left out of the quantitative analysis. Status as participant topic was evaluated using two methods:
- Qualitative assessment of whether the O is directly related to (e.g. is a subpart of) the *(discourse-)unit topic* (UnitTop), in the macro-propositional sense of (2a).
- Quantitative topic-continuity measures. I call this *sequential topic* (SeqTop).

Note that "directly related to unit topic" (UnitTop) and "sequential topic" (SeqTop) have to do with methodological procedures. I am not suggesting these are necessarily distinct statuses in native speaker minds or in cognitive processing.

To illustrate how Os were evaluated by these methods, consider (20). Propositions surrounding the ditransitive clause in (20b) are in free translation. The overall discourse topic, in the macro-propositional sense of (2a), is something like 'What happens when a child is born'. The macro-propositional topic of (20a–c) is 'the business of giving out sugar'. That for (20d) is 'slaughtering the purifying ram'. At its opening, (20d) refers back to the summarizing macro-propositional topic of (20a–c) with the phrase 'the business of sugar'.[10]

(20) a. 'When a woman gives birth,'
   b.
VERB	R		THEME
n-é-ɪshɔr-ɪ	ɪn=kéra	pɔɔkín	ɛ=sʊkári
CN1-3-give-IMPS	FPL=children	all	FSG=sugar

   'all children are given sugar'
   c. 'to put on the hands so they lick it. Every woman is given sugar which she takes to her house.'
   d. 'A castrated ram called "the purifier" is slaughtered when the sun rises (at) the time when the business of sugar is finished, and the shooting of cows [on the jugular vein] has also been done, (then) the castrated ram called "the purifier" is slaughtered. Now this is what this slaughter of the castrated ram called "the purifier" is like...' (birthing, eishoi.001-005)

---

10 As an anonymous reviewer has pointed out, the different "levels" of macro-propositional topics reflect the fact that wide-scope discourse topics can span over multiple narrower-scope discourse topics. There is no limit to such discourse-topic embedding.

In (20b), the R 'all the children' is animate, non-referential, and has zero anaphoric continuity. It has cataphoric continuity into the next two clauses 'to put it on the hands so they lick it'. The non-referential THEME 'sugar' has zero anaphoric continuity, but has cataphoric continuity into the following four clauses, 'to put on the hands so they lick it. Every woman is given sugar which she takes...' Thus, 'children' and 'sugar' are equal in anaphoric continuity, but 'sugar' has greater cataphoric continuity. Also, 'sugar' is an essential part of the local macro-propositional topic 'the business of giving out sugar', whereas 'children' is less so, judged qualitatively. By these criteria, (20) with 'children' before 'sugar' does not conform to a principle like "more topical occurs first".

We now turn to quantified results. Figure 3 presents topic-continuity measures. The anaphoric measure (AnphDist1-2) evaluates whether the referent was mentioned in the first or second immediately preceding clause. The first cataphoric measure (CtphDist1-2) evaluates whether a referent is mentioned in the first or second following clause. The second cataphoric measure (CtphDens3+) evaluates whether a referent is mentioned three or more times in the next ten clauses. When evaluated over all 67 ditransitives, only the anaphoric measure is significant (p = 0.0127, 2-tailed Fisher's Exact Test).

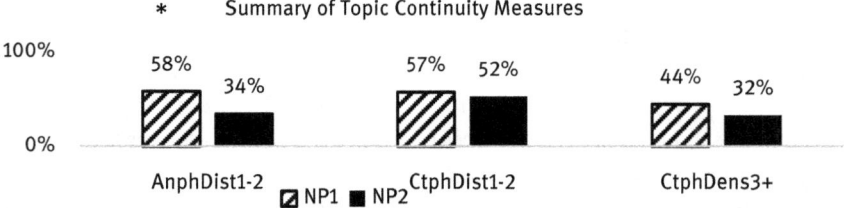

Figure 3: Topic continuity measures for ditransitive Os (N = 67).

Figure 4 presents results for animacy and topic type when the two O NPs differ in the indicated measure. Clauses where both Os were animate or both inanimate were excluded. In the remainder (N=51), 94% of first Os were animate (Anim).

Figure 4 also evaluates whether NP1 versus NP2 is a better *sequential topic*. This is a somewhat higher-level evaluation of the anaphoric and cataphoric measures in Figure 3, and considers only clauses where anaphoric/cataphoric measures were different between the two Os. The number of times NP1 had higher anaphoric and/or cataphoric continuity than NP2 were summed (this sum could be higher than the number of clauses considered) and this was divided by the total number of summed NPs in the subsample. This was compared with same measures for NP2. O referents were also evaluated for match to the *discourse* or *unit topic*, and clauses were excluded if both Os were equally good matches. In the retained clauses (N = 38), 68% of first Os corresponded to (part of) the unit topic, while 32% of second Os did.

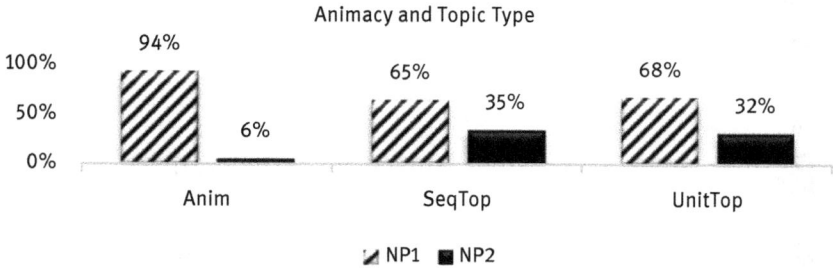

Figure 4: Animacy and topic type of unequal ditransitive Os.

The differences between NP1 and NP2 in Figure 4 are strong for each factor, but animacy is an especially strong correlate of NP1.

We now look at the four ditransitive types individually. Figure 5 examines 'give' ditransitives for the same measures as in Figures 3 and 4. For animacy, sequential topic, and unit topic, only clauses where there is a difference between NP1 and NP2 are considered. Since these parameters can vary independently, the N in each case may differ. Significant topic-continuity measures are anaphoric distance (p = 0.0348) and cataphoric density (p = 0.0092, 2-tailed Fisher's Exact test).

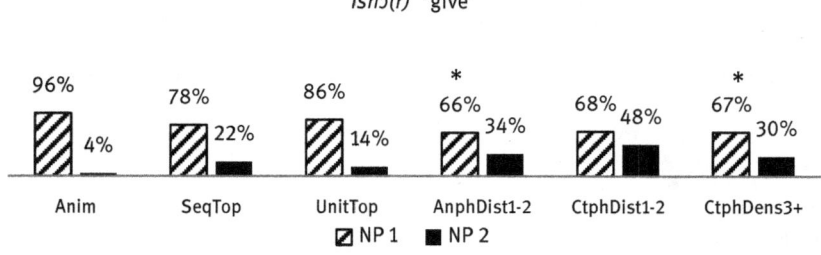

Figure 5: Factors associated with 'give' ditransitive Os (N = 27 Anim; N = 18 SeqTop; N = 22 UnitTop; N = 31 anaphoric & cataphoric measures).

Figures 6, 7, and 8 give analogous information for 'put', instrumental, and dative applicative ditransitives. For each, THEME, R or INSTRUMENT O could be NP1 or NP2. No individual topic-continuity measures are significant for any of these, which might partially be due to the smaller sample sizes. Comparison of the striped and solid bars for topic measures in Figures 5 through 8 suggests that instrumental ditransitives may have a different profile from the other ditransitive constructions.

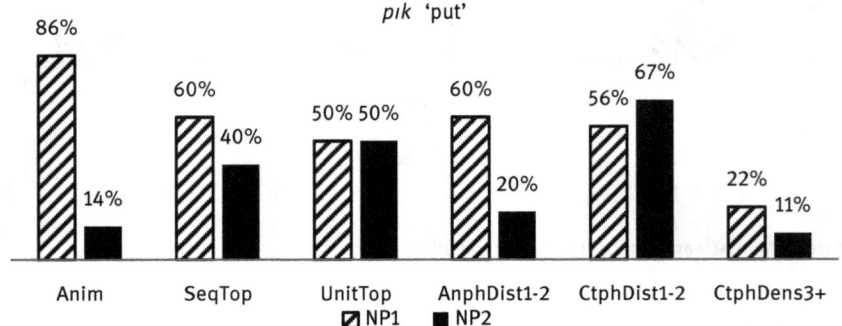

Figure 6: Factors associated with 'put' ditransitive Os (N = 7 Anim; N = 10 SeqTop; N = 4 UnitTop; N = 10 anaphoric & cataphoric measures).

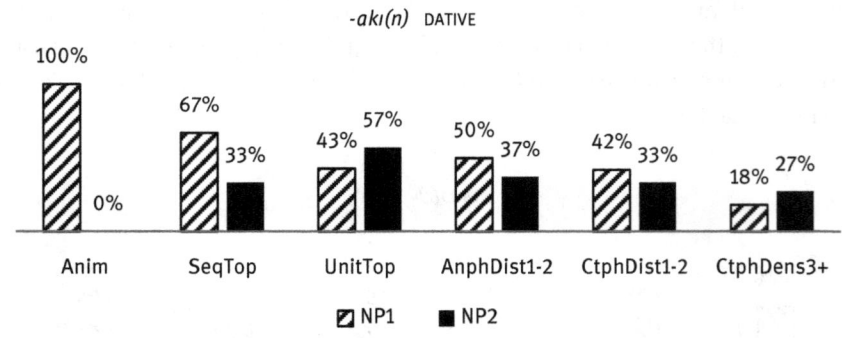

Figure 7: Ditransitives with dative applicative O (N = 11 Anim; N = 15 SeqTop; N = 7 UnitTop; N = 16 anaphoric & cataphoric measures).

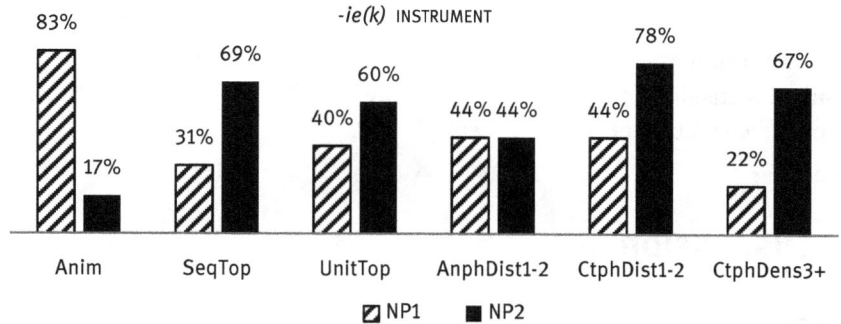

Figure 8: Ditransitives with instrument applicative O (N = 6 Anim; N = 13 SeqTop; N = 5 UnitTop; N = 10 anaphoric & cataphoric measures).

Figures 4 through 8 show that across all ditransitive types, animacy is a paramount correlate of order when there is an animacy disparity between the Os. To probe what might be operative when Os do not differ in animacy, Figure 9 evaluates animate-animate and inanimate-inanimate clauses for the other factors. Note that the number of animate-animate clauses is especially small.

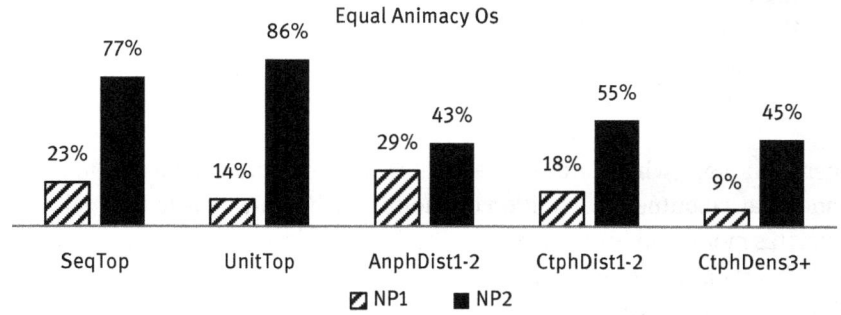

Figure 9: Equal-animacy Os (Total N = 13; 4 an-an; 9 inan-inan).

Figures 8 and 9 suggest a reversal of the topicality relationship between NP1 and NP2, compared to Figures 3–7. However, it is difficult to know what to make of this especially for Figure 9, given: 1. the small equal-animacy numbers, 2. the fact that there is a potentially differing topicality pattern for the instrumental, 3. some examples involve pronouns, and 4. some may involve focus subtypes. In

(11) above, for example, the strange visitors say that their age-set does not drink tea with cups, and command their tea to be served in a stool. In (11a), 'we drink tea' is presupposed while 'with cups' is under negative focus. Nevertheless, marked focus does not necessarily correlate with NP1 position: in (11b), 'seats' contrast with 'cups' of (11a); but 'seats' is NP2.

5 Discussion and conclusions

Keeping in mind that some sample sizes are small, the major results are:
1. Order does not correspond to semantic (macro-)role like R or THEME.
2. When there is an animacy difference, NP1 is overwhelmingly animate. A relationship between animacy and R likely accounts for the predominance of V-R-THEME orders with 'give' in Figure 1.
3. Free pronouns are dominantly NP1.
4. Overall, there is little correlation between order and *unit topic*.
5. Overall, there is no clear association between order and *sequential topic*. There is some correlation in 'give' clauses, but this could be confounded by animacy.
6. Different ditransitive constructions may have different profiles. Disregarding animacy, topicality factors affecting postverbal O order in instrument ditransitives may pattern distinctly from the other ditransitive types. There may be differences between 'put' and 'give' constructions, though this may be confounded by animacy.

There is a somewhat natural relationship between animacy and topic status under one "aboutness" definition or another, as humans may tend to talk about animates more than inanimates. Result 3 might reflect topicality: normally, only participants already established in discourse are referred to with definite (and phonologically light) pronoun forms, and we tend to talk "about" established ("topical") participants. Troubling for an "NP1 = Topic" hypothesis, however, are results 4 and 5. These concern notions of topic in the information-structure literature, but they throw doubt on a relationship strictly between "topic" and position, at least across the ditransitive types taken together. More factors must be at play than what the topic measures alone reveal.

Finally, though the association between animacy and NP1 is especially strong, a simple principle of "animate first" does not suffice, as the correlation is not 100% and doesn't account for equal-animacy situations.

The findings here are largely congruent with Payne et al. (1994), which examined order of postverbal subject and object. That study used the same topic-continuity methodology as here, but did not look at animacy or *unit topic*. It found that linearly first arguments after the verb in both VSO and VOS clauses were anaphorically more continuous than linearly second arguments. That study also found that NP1 arguments were more cataphorically continuous than NP2 (Payne et al. 1994: 309). If either subject or object was expressed by a free pronoun and the other argument by a lexical NP, the pronoun consistently occurred next to the verb (Payne et al. 1994: 306). Though this generally would accord with a "phonologically light before heavy" principle, we have pointed out instances that do not follow such a principle.

Given the patterns of what correlates with NP1 position in both the ditransitive O data and in VSO versus VOS clauses, I suggest that a broader cognitive notion than just animacy or phonological weight is relevant to what occurs first, something more akin to proximal status as characterized in (4) above. "Proximal" is understood here as the participant in the clause that is more familiar and/or closer to ego in the interlocutor's cognitive discourse world regardless of animacy or humanness per se, compared to another cognitively "distal" participant. The higher frequencies of V-R-THEME (Figure 1) and VSO (91% in Payne et al. 1994), compared to V-THEME-R and VOS, result from how linguistic principles map what is cognitively proximal to syntactic subject and to the semantic macro-role of R. The cognitively proximal argument might – or might not – be a *sequential* or *unit topic* in some instances. If this is right, the basic Maa information structure construction can be represented as in Figure 10.

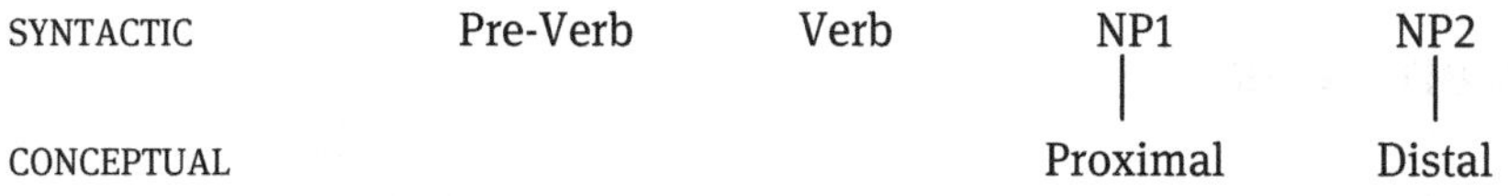

SYNTACTIC	Pre-Verb	Verb	NP1	NP2
CONCEPTUAL			Proximal	Distal

Figure 10: The Maa basic information structure construction.

Finally, it is true that we have not examined all the factors in (4) above that might pertain to cognitive proximal versus distal status, and it would be of interest to experimentally evaluate such factors as physical distance, size, and other saliency factors on order of postverbal phrases, especially in the equal-animacy situations. Additionally, whether there is a unifying conceptual status for all pre-verb phrases remains for further study.

Abbreviations

1, 2, 3	1st, 2nd, 3rd person
CN	connective
DAT	dative
F	feminine
FPL	feminine plural
FSG	feminine singular
IMP	imperative
IMPS	impersonal
INCHO	inchoative
INF	infinitive
INST	instrument
M	masculine
MSG	masculine singular
MPL	masculine plural
NEG	negative
NOM	nominative
OBL	oblique
OPT	optative
PF	perfect(ive)
PL	plural
PSD	possessed
PROG	progressive
REL	relativizer
SG	singular
SBJV	subjunctive

References

Allan, Keith. 1990. Discourse stratagems in a Maasai story. In John P. Hutchinson & Victor Manfredi (eds.), *Current approaches to African linguistics* 7, 179–191. Dordrecht: Foris.

Arnold, Jennifer, Thomas Wasow, Anthony Losongo & Ryan Ginstrom. 2000. Heaviness vs. newness: The effects of structural complexity and discourse status on constituent ordering. *Language* 76. 28–55.

Chafe, Wallace. 1976. Givenness, contrastiveness, definiteness, subjects, topics, and point of view. In Charles Li (ed.), *Subject and topic*, 25–55. New York: Academic Press.

Chafe, Wallace. 1987. Cognitive constraints on information flow. In Russel Tomlin (ed.), *Coherence and grounding in discourse*, 21–52. Amsterdam & Philadelphia: John Benjamins.

DeLancey, Scott. 1981. An interpretation of split ergativity and related patterns. *Language* 57. 626–657.

Dik, Simon, Maria E. Hoffmann, Jan R. de Jong, Sie Ing. Djiang, Harry Stroomer & Lourens de Vries. 1981. On the typology of focus phenomena. In Teun Hoekstra, Harry van der Hulst & Michael Moortgat (eds.), *Perspectives on functional grammar,* 41–74. Dordrecht: Foris.

Dryer, Matthew. 1992. A comparison of the obviation systems of Kutenai and Algonquian. In Cowan, William (ed.), *Papers of the Twenty-Third Algonquian Conference*, 119–163. Ottawa: Carleton University.

Fauconnier, Gilles. 1994. *Mental spaces: Aspects of meaning construction in natural language.* New York: Cambridge University Press.

Firbas, Jan. 1964. On defining the theme in functional sentence perspective. *Travaux Linguistiques de Prague* 1. 267–280

Gernsbacher, Morton. 1995. The structure building framework: What it is, what it might also be, and why. In Bruce K. Britton & Arthur C. Graesser (eds.), *Models of text understanding,* 289–311. Hillsdale, NJ: Erlbaum.

Givón, Talmy (ed.). 1983. *Topic continuity in discourse: A quantitative cross language study.* Amsterdam & Philadelphia: John Benjamins.

Gundel, Jeanette, Nancy Hedberg & Ron Zacharski. 1993. Cognitive status and the form of referring expressions in discourse. *Language* 69. 274–307.

Hockett, Charles. 1958. *A course in modern linguistics.* New York: The Macmillan Company.

Hollis, Alfred Claud. 1905. *The Masai, their language and folklore.* Oxford: Clarendon Press.

Jacobs, Joachim. 2001. The dimensions of topic-comment. *Linguistics* 39. 641–681.

Koopman, Hilda. 2005. On the parallelism of DPs and clauses: Evidence from Kisonko Maasai. In Andrew Carnie, Heidi Harley & Sheila Ann Dooley (eds.), *Verb first: On the syntax of verb initial languages,* 281–301. Amsterdam & Philadelphia: John Benjamins.

Korta, Kepa & John Perry. 2020 (Spring). Pragmatics. In Edward N. Zalta (ed.), *The Stanford encyclopedia of philosophy.* Stanford: Stanford University. https://plato.stanford.edu/archives/spr2020/entries/pragmatics/

Lambrecht, Knud. 1994. *Information structure and sentence form: Topic, focus, and the mental representation of discourse referents.* Cambridge: Cambridge University Press.

Lewis, M. Paul, Gary F. Simons & Charles D. Fennig (eds.). 2014. *Ethnologue: Languages of the world.* 17th edn. Dallas, Texas: SIL International. Online version: http://www.ethnologue.com.

Matić, Dejan & Daniel Wedgwood. 2013. The meanings of focus: The significance of an interpretation-based category in cross-linguistic analysis. *Journal of Linguistics* 49. 127–163.

Myachykov, Andriy. 2007. *Integrating perceptual, semantic, and syntactic information in sentence production.* University of Glasgow PhD dissertation.

Payne, Doris L. (ed.). 1992. *Pragmatics of word order flexibility.* Amsterdam & Philadelphia: John Benjamins.

Payne, Doris L. 1995. Verb initial languages and information order. In Pamela Downing & Michael Noonan (eds.), *Word order in discourse,* 449–485. Amsterdam & Philadelphia: John Benjamins.

Payne, Doris L. 1999. What counts as explanation? A functionalist approach to word order. In Michael Darnell, Edith Moravcsik, Frederick J. Newmeyer, Michael Noonan & Kathleen Wheatley (eds.), *Functionalism and formalism in linguistics, vol. 1,* 135–163. Amsterdam & Philadelphia: John Benjamins.

Payne, Doris L. 2011. The Maa (Eastern Nilotic) impersonal construction. In Anna Siewierska & Andrej Malchukov (eds.), *Impersonal constructions. A cross-linguistic perspective,* 257–284. Amsterdam & Philadelphia: John Benjamins.

Payne, Doris L. 2015. Topic versus proximate: Information order in Maa (Eastern Nilotic). In Hieda Osamu (ed.), *Information structure and Nilotic languages*, 37–59. Tokyo: Research Institute for Languages and Cultures of Asia and Africa, Tokyo University of Foreign Studies.

Payne, Doris L., Mitsuyo Hamaya & Peter Jacobs. 1994. Active, inverse, and passive in Maasai. In Talmy Givón (ed.), *Voice and inversion*, 283–315. Amsterdam & Philadelphia: John Benjamins.

Posner, Michael I. & Gregory J. DiGirolamo. 1998. Executive attention: Conflict, target detection and cognitive control. In Raja Parasuraman (ed.), *The attentive brain*, 401–423. Cambridge, MA: MIT Press.

Prince, Ellen. 1981. Toward a taxonomy of given-new information. In Peter Cole (ed.), *Radical pragmatics*, 223–256. New York: Academic Press.

Reinhart, Tanja. 1982. Pragmatics and linguistics. An analysis of sentence topics. *Philosophica* 27. 53–94.

Rooth, Mats. 1992. A theory of focus interpretation. *Natural Language Semantics* 1. 75–116.

Stalnaker, Robert. 1999. *Context and content*. Oxford: Oxford University Press.

Stalnaker, Robert. 2002. Common ground. *Linguistics and Philosophy* 25. 701–721.

Tomlin, Russell. 1995. Focal attention, voice, and word order: An experimental, cross-linguistic study. In Pamela Downing & Michael Noonan (eds.), *Word order in discourse*, 517–554. Amsterdam & Philadelphia: John Benjamins.

Tucker, Archibald N. & John T. Mpaayei. 1955. *Maasai grammar, with vocabulary*. London: Longman, Greens.

Vallduví, Enric & Maria Vilkuna. 1998. On rheme and kontrast. In Peter Culicover & Louise McNally (eds.), *The limits of syntax*, 29–106. New York: Academic Press.

Vallejos, Rosa. 2009. The focus function(s) of *=pura* in Kokama-Kokamilla discourse. *International Journal of American Linguistics* 75. 399–432.

van Dijk, Teun. 1977. Sentence topic and discourse topic. *Papers in Slavic Philology* 1. 49–61.

Watters, John. 1979. Focus in Aghem: A study of its formal correlates and typology. In Larry M. Hyman (ed.), *Aghem grammatical structure*, 137–197. Los Angeles: University of Southern California.

Zúñiga, Fernando. 2006. *Deixis and alignment: Inverse systems in indigenous languages of the Americas*. Amsterdam & Philadelphia: John Benjamins.

Thomas Jügel
Word order variation in Middle Iranic: Persian, Parthian, Bactrian, and Sogdian

Abstract: This paper focuses on word order structures that include Targets (i.e., Goals, Recipients, Addressees, etc.) in four Middle Iranic languages (Bactrian, Middle Persian, Parthian, and Sogdian) by querying the ergative database (Jügel 2015) and examining the *Vessantara Jātaka* (Benveniste 1946). Targets can appear in various positions, including the postverbal position, which is prevalent in a number of modern Iranic languages. The paper seeks to identify possible triggers for the variation, such as morphosyntactic ones as well as those related to information structure.

Keywords: focus; language change; Middle Iranic/Middle Iranian; postverbal Target; topic

1 Introduction

Iranic languages[1] are commonly considered languages in which the finite verb appears in the final position.[2] However, it has been noted that a specific group of participants that is characterized as the endpoint of a motion tends to appear in the immediate postverbal position if they are neither the subject nor object (cf. ex. 1). Following Asadpour (this volume), this paper subsumes these participants under the term 'Target' (T, set bold in examples), which includes Goals of motion and caused-motion verbs, Addressees of *verba dicendi* (verbs of utterance), Recipients of verbs of transfer, and partly also Experiencers and Beneficiaries.

---

[1] The adjective 'Iranic' was chosen to refer to Irano-Aryan languages over the more commonly used 'Iranian' in order to avoid ambiguity with references to the state of Iran (cf. Schmitt 1989: §1.1.3).
[2] For Old Iranic see Skjærvø (2009: 94 §5.1), who speaks of "basic word order [...]: SOV" (i.e., subject – object – verb). All Iranic languages in Dryer (2013) are also classified SOV. Jügel (2015: 394) shows that SOV is the most frequent but not the prevalent order in Bactrian (40%), Parthian (39%), and Middle Persian (56%).

Thomas Jügel, Center for Religious Studies, Ruhr-University Bochum, Universitätsstraße 90a, 44789 Bochum, Germany

https://doi.org/10.1515/9783110790368-003

(1) **SOVT** (Middle Persian, *Šāyest nē šāyest* 15 §30)[3]　　[Kotwal 1969: 66]
　　　ohrmazd　　　*ēn=iz*　　　*guft*　　　*ō*　　　***zarduxšt***
　　　PN　　　　　　this=ADD　　say.PST.3SG　to　　PN
　　　'Ohrmazd also said this to Zarduxšt [that ...]'

Studying the position of Goals[4] in languages of East Anatolia (including Northern Kurdish), Haig (2017: 410) stipulates a "Mesopotamian OVG word order". Haig and Khan (2019: 24) call "northern Iraq and neighbouring regions of western Iran" the apparent "areal epicentre". This could be interpreted that this region would be a geographical starting point of "OVG" word order in Iranic languages, where – according to Haig and Khan (2018: 23) – this order is said to be "disharmonic". If so, Iranic languages would have copied this structure from other languages.

This study tests whether this OVG order (OVT following our terminology) is a new phenomenon in Iranic languages and whether there is evidence for a geographical rule that would support the idea of an epicenter. It does not attempt to establish a rule-based grammar of word order patterns for Middle Iranic in general. Information on West-Iranic Parthian, Middle Persian and East-Iranic Bactrian are available in the database of Middle Iranic ergative structures (Jügel 2015: 29f.).[5] For East-Iranic Sogdian, this paper uses the Sogdian version of the *Vessantara Jātaka* (Benveniste 1946). For an approximate position of these four languages and their geographical distance to the assumed Mesopotamian epicenter of OVT, see Map 1.

The database of Middle Iranic ergative structures only contains finite past tense verb forms with grammatical information on subject and object. However, it was possible to retrieve Targets flagged by the preposition 'to, towards' (Middle Persian/Parthian *ō*, Bactrian *av*)[6], the position of which could be identified in relation to the verb.

---

3 Glossing of all examples added by myself.
4 Haig (2017: 408) uses 'Goal' to refer to "recipients, addressees as well as local goals" encoded by "full NPs".
5 For Bactrian, I noted 35 clauses with Targets flagged by 'to' out of 503 clauses altogether; for Parthian 106 clauses with Targets out of 1,313 clauses; for Middle Persian 978 clauses with Targets out of 12,548 clauses.
6 It is striking that the relative number of constituents flagged by 'to' is nearly the same for all three languages (7–8.5%).

Sogdian, on the other hand, was checked against the complete text of the Sogdian *Vessantara Jātaka*, a Sogdian re-narration of the Sanskrit version.[7] The risk is low that it exhibits features of a translated text, and every kind of Target construction could be taken into account for Sogdian irrespective of morphological marking.[8]

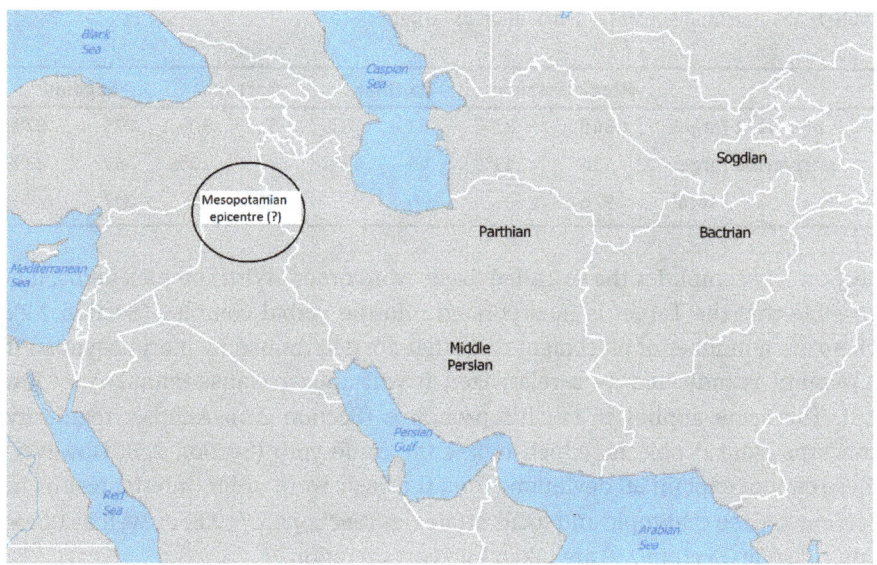

Map 1: Approximate location of the investigated Middle Iranic languages.

2 Survey

This survey examines variation of word ordering in four Middle Iranic languages. Table 1 gives an overview of Target positions in relation to the finite verb (pre- and postverbal). Apparently both positions are possible, but the preverbal position is clearly preferred. Only Parthian shows a fairly high percentage of postverbal Targets, which is probably due to characteristics of the Parthian corpus (cf. Section 2.9).

---

[7] One passage practically repeats the same text ten times, namely the prince's encounter with ten Brahmins during his journey into exile. This survey only analyses the first encounter.
[8] For Sogdian, 317 clauses with Targets out of more than 1,300 sentences were noted.

The ergative database query also shows that there are no apparent constraints of word orders in main or subordinate clauses. However, main clauses show more variation than subordinate clauses. Postverbal position of Targets is not excluded in subordinate clauses, but appears less often than in main clauses.

Table 1: Absolute and percental distribution of Targets.

	Middle Persian		Parthian		Bactrian		Sogdian	
preverbal Target	848	87%	61	58%	28	80%	275	87%
postverbal Target	130	13%	45	42%	7	20%	42	13%
Total	978		106		35		317	

Section 2.1 exemplifies the so-called 'basic word order'. With complex verbs, it is possible that the Target is incorporated into the verbal complex (Section 2.2). There are a number of placement rules that are determined by morphosyntax. If a referent is indicated by a relativizer, it will appear clause-initially (Section 2.3). The same applies to enclitic pronouns (Section 2.4). Another regularity concerns clausal objects, which follow the finite verb (Section 2.5). However, these cannot explain all deviations from the basic word order (interim results in Section 2.6), so that topic and focus strategies (Sections 2.7–2.8) as well as questions of style (Section 2.9) are taken into consideration.

2.1 'Basic word order'

As mentioned in the introduction, Iranic languages are usually classified as V-final languages with the structure SOV. In this so-called 'basic word order', the Target can appear before the object (TOV) or after it (OTV), cf. ex. (2)–(5). Middle Persian prefers the order OT, while Sogdian shows a few more cases of TO. The Parthian and Bactrian numbers are very low (see Table 2). Based on the available information, it is not possible to substantiate one sequence or the other.

Table 2: Object-Target sequence.

	Middle Persian	Sogdian	Parthian	Bactrian
TOV	35	48	7	2
OTV	188	40	8	2

(2) **STOV** (Parthian, M48 II R.27–28 – V.1[9])
[Sundermann 1981: 21–22, ll. 63–65)][10]

aδyān	mārmānī	ō	**tūrān**	**(šā)[h]**
then	PN	to	PN	king

(wa)s	[xra](d	u)[d]	žī[rī](ft)	wifrāšt
much	wisdom	and	sagacity	teach.PST

'Then Mār Mānī taught much wisdom and sagacity to the Tūrān king.'

(3) **STOV**[11] (Middle Persian, *Kārnāmag ī Ardaxšīr* 14 §2) [Čunakova 1987: 54]

u=šān	ō	**xwahar**	**ī**	**xwēš**	čiyōn	zan	ī	ardaxšīr
and=PC.3PL	to	sister	EZ	own	how	woman	EZ	PN

būd	nāmag	nibišt
be.PST	letter	write.PST

'And they wrote a letter to their sister as (she) was the wife of Ardaxšīr.'

(4) **SOTV** (Sogdian, *Vessantara Jātaka* ll. 200–202) [Benveniste 1946: 16]

pārti	vay-a	əxu	wispəšē ō	wxušu	ənsur
but	sir-VOC.SG	ART.NOM.SG.M	prince ART.ACC.SG.M	six	tusk

əspētē	rājvart	pīδ-ān	xutāw	əwī	**brāman-t**
white	PN	elephant-GEN.PL	king	ART.DAT	Brahmin-PL

ϑvār	ϑvart=δār-t
gift	give.PP=PRF-3SG

'But, sir, the prince has given the white Rājvart with six tusks, king of the elephants, to the Brahmins as a gift!'

(5) **SOTV** (Bactrian, df 8–9) [Sims-Williams 2007: 107][12]

(οτ)ομο	λασο	ι	β(ρ)ακο	παμα(νο)[.]
ut=əm	las	i	vr̃ak	pāman[...]
and=PC.1SG	ten	EZ	?	wool-?

αβο	**το**	**βραδο**	φοþ(ταδο)
av	**tu**	**vrād**	**fəštād**
to	you	brother	send.PST

'And I have sent ten ... of wool(?) ... to you, my brother.'
[Sims-Williams 2007: 106]

---

9 Cf. http://turfan.bbaw.de/dta/m/images/m0048plus_seite2_detail2.jpg.
10 This is the only instance of a simplex word in STOV order, which unfortunately appears in a damaged passage. However, word order can be safely established, because only the object is barely legible.
11 Similar to a relative clause, the causal subclause gives information on the Target referent.
12 Although the object of this example is not fully legible, the word order can still be established.

2.2 Incorporated position

Asadpour (this volume) includes an "incorporated position" in his analysis of Target ordering. This refers to the common phenomenon of complex verbs/predicates (a.k.a. 'light verb construction') in Iranic languages. These complex verbs consist of at least two elements: the preverb and the 'light verb', which form one semantic unit.[13] If the preverb derives from a noun, it may hand its complement on to the newly formed complex verb. This complement can continue to be linked to the preverb by nominal morphology. As a result, it appears 'inside' the verbal complex, viz. following the preverb. The following examples illustrate constructions that probably served as precursors of such structures, e.g., *menišn ō ... nihād* 'to give thought to ...' or 'to decide to ...' (ex. 6). Prepositional complements of nouns usually follow their head (*menišn ō ...*). When the noun is reinterpreted as the preverb, the nominal complement becomes a verbal complement (*menišn nihād + ō ...*), but still appears in its original position. This explains the position of the Target following the object (SOTV), or preverb (SVTV) in ex. (6). In ex. (7), *abāz* 'back' could be considered a specification of the preposition *ō* 'to' or as a preverb, i.e., *abāz xwāst* 'call back' (cf. German '(zu einem Ort) zurückrufen').

(6) **SOTV/SVTV** (Middle Persian, *Kārnāmag ī Ardaxšīr* 3 §9) [Čunakova 1987: 4]
   ardaxšīr ... menišn   ō   **wirēxtan**   az   ānōh   nihād
   PN   thinking to flee.INF   from there put.PST
   'Ardaxšīr (when he heard the news) gave thought to fleeing from there.'

(7) **SOTV/SOVTV** (Middle Persian, *Kārnāmag ī Ardaxšīr* 10 §6)
                                                           [Čunakova 1987: 49]
   haftōbād spāh   ī   xwēš hāmōyēn   abāz ō   **dar**   xwāst
   PN   army   EZ   own   altogether back to   court   call.PST
   'Haftōbād called his whole army back to (his) court.'

2.3 Structures with relativizers

Relative clauses begin with the relativizer in the four Middle Iranic languages. The relativizers originate from relative or interrogative pronouns and develop to relative particles or subordinators during the Middle Iranic period. The antecedent of the relative clause can then be omitted or resumed by another pronoun.

---

13 See Samvelian (2012) for details.

This means that it is uncertain whether the relativizer is a pronoun that encodes the subject, object, or Target, or whether the participant is omitted and the relativizer, as a particle, is irrelevant for word order classification. The constituent that is either encoded by a relative pronoun or omitted is put in parentheses.

(8) **(S)TOV** (Sogdian, *Vessantara Jātaka* ll. 337–338) [Benveniste 1946: 22]
xunāx	mayδəv-i	əkēti	**əwēn**	**suδāšan**
that.NOM.SG.M	minister-NOM.SG.M	REL	ART.DAT.SG.M	PN
ō	əžwān	ϑāvar		
ART.ACC.SG.M	life	give.IPRF.3SG		

'that minister who saved Suδāšan's life [*lit.* gave the life to Suδāšan]'

(9) **(S)OTV** (Middle Persian, *Kārnāmag ī Ardaxšīr* 15 §19) [Čunakova 1987: 58]
andar	ham	zamān	kas	mad
in	same	time	someone	come.PST
kē	šābuhr	ō	**ānōh**	nīd
REL	PN	to	there	lead.PST

'At the same time, someone came who brought Šābuhr there.'

(10) **(O)STV** (Sogdian, *Vessantara Jātaka* ll. 1475–1476) [Benveniste 1946: 84]
yunē	kavnak	wiδvāy	čuti əzu **təwa**	nūr	əkϑār-ām
this.NOM.SG.M	little	exposition	REL I you.OBL	today	do.PRF-1SG

'this little exposition, which I have given you today'

If the antecedent of the relativizer is coreferential with the subject of the relative clause, the Target usually appears in preverbal position (58× Middle Persian, 8× Parthian, 3× Sogdian) and rarely in postverbal position (5× Middle Persian, 3× Parthian). If the antecedent is coreferential with the object, the Target mostly appears in preverbal position as well (29× Middle Persian, 2× Parthian, 4× Bactrian, 2× Sogdian vs. postverbal position 3× Middle Persian, 4× Parthian).

In a few cases, the relativizer's status as a particle or subordinator is apparent, because the antecedent is resumed by another pronoun in the relative clause (all preverbal: 10× Middle Persian, 1× Parthian, 1× Bactrian). However, in at least one case, the relativizer is a pronoun (Parthian *ō kū* 'whereto').

2.4 Structures with enclitic pronouns

Enclitic pronouns are only used for oblique constituents. Thus, a Target can be encoded by an enclitic pronoun in all four Middle Iranic languages. In an ergative construction, the subject is in oblique case in contrast to an accusative construction where this is the object. Since my corpus of Middle Persian, Parthian and Bactrian only contains ergative constructions, only subjects and Targets can be encoded by enclitic pronouns. There are very few ergative constructions in the Sogdian corpus, which explains the lack of oblique subjects for Sogdian.

If one of the constituents is encoded by an enclitic pronoun, it usually attaches to the first word of the clause that carries stress (cf. ex. 11).

(11)   **STOV** (Bactrian, J 4)   [Sims-Williams 2012: 49]
   οτηνο   αβο   μασκο   μολρο   ταβδο
   ut=ēn   *av*   *mask*   *mulr*   *tavd*
   and=PC.3PL to   hereupon   seal   impress.PST
   'And they have placed (their) seals hereupon.'

The following tables give an overview of various word orders when either subject or Target are encoded by an enclitic pronoun. Encoding of the subject by enclitic pronouns is not attested in Sogdian, the corpus of which exhibits examples for objects instead (Sogdian: 14× OTV, 3× OVT, 3× OSVT, 2× OSVT, see ex. 12).

(12)   **OSTV** (Sogdian, *Vessantara Jātaka* ll. 1208–1209)   [Benveniste 1946: 72]
   əti=šu   *təyu*   *āδē*   *ϑvār*   *nē*   *ϑvār-ē*
   that=PC.3SG.ACC   you.NOM.SG   someone   gift   NEG   give-2SG
   'that you do not give her to anyone as a gift'

Table 3: Subject = enclitic pronoun.

		Middle Persian	Parthian	Bactrian
	STV	34	3	3
	SOTV	100	4	1
preverbal T	OSTV	14	–	4
	STOV	15	3	2
	STVO	29	2	2
	SVT	1	–	–
	SOVT	1	3	–
	OSVT	3	3	–
postverbal T	SVTO	12	–	3
	VST	1	–	–
	VSTO	6	–	–
	SVOT	4	–	–
	VSOT	–	4	–

(13) **SOTV** (Parthian, M4575 R.ii.1–6)   [Sundermann 1981: 56–57, ll.657–659]
 aδyān=iš   (pa)ttēg   masādar   aδ   hannī   brādar
 then=PC.3SG PN   presbyter with   PN   brother
 ō   hīndūgān   ō   dēb   frašud
 to   India   to   PN   send.PST
 'Then he sent the presbyter Pattikios together with brother Ḥannī to India to (the city of) Dēb.'

(14) **SOTV** (Bactrian, df 8–9)   [Sims-Williams 2007: 107][14]
(=5) (οτ)ομο   λασο   ι   β(ρ)ακο   παμα(νο)[.   ]
 ut=əm   las   i   vrăk   pāman[...   ]
 and=PC.1SG   ten   EZ   ?   wool-?
 αβο   το   βραδο   φοþ(ταδο)
 av   tu   vrād   fəštād
 to   you   brother   send.PST
 'And I have sent ten ... of wool(?) ... to you, my brother.'
                                              [Sims-Williams 2007: 106]

Table 4: Target = enclitic pronoun.

		Middle Persian	Parthian	Bactrian	Sogdian
	TV	–	–	1	1
	TVO	–	–	–	33
	TOV	1	–	–	3
	TVS	–	–	–	1
preverbal T	TSV	–	4	–	3
	TVS	–	–	1	–
	TSOV	–	–	–	30
	TSVO	–	5	–	21
	TVSO	–	–	–	2
postverbal T	SOVT	–	–	1	–

(15) **TSVO** (Sogdian, *Vessantara Jātaka* ll. 5–6)   [Benveniste 1946: 2]
 rəti=šī   xa   xutēn   māδ   pətīškway
 and=PC.3SG   ART.NOM.SG.F   queen   thus   say.IPRF.3SG

---

14 Although the object is not fully legible in this example, there is enough information to establish the word order.

mā𝛳	*əti*	*vay-a*	*men-u*
thus	DS	sir-VOC	think-IPRF.1SG

'And the queen said to him thus: Thus, sir, (I) thought ...'

In Middle Persian, the appearance of enclitic Targets in the initial position leads to ambiguity, because the governing adposition usually remains *in situ* with a placeholder attached (*awi=š*, cf. ex. 16).[15] It is unclear whether the semantic nucleus (i.e., the enclitic pronoun) or the position of flagging (i.e., the adposition together with the placeholder) should be conclusive for word order classification: TSV or STV. Table 5 illustrates this difference.

(16) **TS(T)V** (Middle Persian, *Abar ēwēnag nāmag nibēsišnīh* 20)

[ʿOryān 2001: 310]

ka=mān	*āgāhīh*	[...]	***awi=š***	*mad*
when=PC.1PL	news		to=PC.3SG	come.PST.3SG

'when the news reached us' (*lit.* 'when us the news thereto reached')

Table 5: Taking the complement (PC) or the adposition (*awiš*) as T.

	STV	TSV	SVT	TVS	TV
T = PC	5	15	–	1	4
T = *awiš*	19	–	1	1	4

The word order TSV is only attested in Middle Persian if the enclitic pronoun is considered for the classification. If, instead, the adposition together with the placeholder is taken as the position of T, the 15× TSV orders are re-classified as 14× STV and 1× SVT orders.

2.5 Structures with complement clauses

If one of the constituents is a complement clause, it is always the last element of the order, and most word orders with final objects refer to structures with complement clauses.

---

15 On this placeholder construction, see Jügel (2015: 242–243) and Jügel (2016).

Table 6: Object = complement clause.

		Middle Persian	Parthian	Bactrian	Sogdian
preverbal T	(S)TVO	75	8	11	28
	TSVO	–	–	–	49
	TVSO	–	–	–	1
postverbal T	(S)VTO	10	6	1	1
	VSTO	10[16]	1	2	–
	VTSO	–	1	–	–

(17) **STVO** (Middle Persian, *Kārnāmag ī Ardaxšīr* 17 §14)   [Čunakova 1987: 61]

u=š	ō	**kanīzag**	guft	kū	tō	čē	dān-ē
and=PC.3SG	to	girl	say.PST	that	you.SG	what	know.PRS-2SG

kū	man	šābuhr	h-am
that	I	PN	be.PRS-1SG

'And he said to the girl: How do you know that I am Šābuhr?'

2.6 Interim results

Even if the 'basic word order' is disregarded as well as those where deviation can be explained by the special rules of clitic placement, relativization and complement clauses,[17] a considerable degree of variation still remains (marked grey in Table 7).

The smallest corpus, Bactrian, shows the least variation (with a total of 35 Target structures). Considerable variation is attested for Middle Persian (978 Targets), Sogdian (317 Targets), and especially Parthian (106 Targets). The high degree of variation in Parthian is probably due to the large number of hymns, which are prone to peculiar word ordering.

---

16 In the *Mēnōg ī Xrad* the same sentence with VSTO word order is repeated 61 times.
17 This refers to the following word orders: STOV, SOTV, and OSTV, TSOV (if O or T are encoded by enclitic pronouns or relativizers), and STVO (if O is a complement clause), and combinations of both.

Table 7: "Non-basic" word order.

		Middle Persian	Parthian	Bactrian	Sogdian
preverbal T	OSTV	■	■	■	■
	TSV	■	■		■
	(S)TVO	■	■	■	
	OTSV	■		■	■
	TVS	■		■	■
	TSOV	■	■	■	■
	TSVO	■	■	■	■
	TVSO	■	■	■	■
postverbal T	(S)VTO	■	■	■	■
	(S)VT	■	■	■	■
	(S)OVT	■	■		■
	OSVT	■	■	■	■
	(S)VOT	■	■	■	■
	VST	■	■	■	■
	VTS	■	■	■	■
	VSTO	■	■		■
	VOST	■			■
	VSOT	■			■
	VTSO	■			■
	OVTS	■			■

The remaining cases suggest that other factors determine word order under certain conditions. Jügel (2015: 269–271) presents 11 examples of Bactrian, Parthian, and Middle Persian, in which length seems to be the relevant factor for a constituent to appear in final position. Sometimes, a correlative pronoun can be found *in situ*. For Sogdian see ex. (18), (19), (32), (34), and (41).

(18) **OSVT** (Sogdian, *Vessantara Jātaka* ll. 1303–1306)   [Benveniste 1946: 76]
   rəti=šan   vay-a   əxu   brāman   par   pərāδən
   and=PC.3PL sir-VOC ART.NOM.SG.M   Brahmin   for   selling
   əškar-ti=skun   čann   nāv   əku   nāv=ti   čann   ōtāk
   lead.PRS-3SG=PROG   from   people to   people=and   from   region

əku	ōtāk=ti	čann	kanθ	əku	kanθ
to	region=and	from	city	to	city

'And, sir, the Brahmin is leading them for sale from people to people and from region to region and from city to city.'

Ex. (19) is exceptional, because the phrase 'pay homage' appears in the *Vessantara Jātaka* 18 times, but only four times with the object 'homage' in postverbal position (see also ex. 33). The Target (marked bold) likely appears at the end due to its length.

(19) **SVOT** (Sogdian, *Vessantara Jātaka* ll. 1431–1433) [Benveniste 1946: 82]

rəti	suδāšan	var-a	ənzānūka	nəmāčyu	čann	δūr-ē
and	PN	carry-3SG.IPRF	on_knees	homage	from	far-OBL
əwēn	**ptər=ti**	**əwī**		**māt-ē**		**farn**
ART.DAT.SG.M	father=and	ART.DAT.SG.F		mother-OBL		glory

'And from far Suδāšan paid homage on his knees to the glory of (his) father and mother.'

Beside a constituent's length, its topicality (Section 2.7), its focality (Section 2.8), and questions of style (Section 2.9) may influence its position in a clause.

2.7 Topic strategies

There are few TSOV cases with orthotone Targets, which seem to represent instances of topic setting or topic continuation ('topic' in the sense of aboutness-topic, cf. Lambrecht 1994: 118).

In ex. (20), Suδāšan starts a new conversation with this sentence and explains his plan to his host. The Target of 'to order' is placed clause-initially and the object is an infinitive clause.[18] The Target referent is then continued as the subject of the following intransitive verb and appears again clause-initially as causer ('*I* will go so that *by me* the glory of (my) father will not be harmed'). This suggests that positioning elements clause-initially is a strategy for topic setting and topic continuation.[19]

---

[18] The matrix verb usually precedes the infinitive verb and thereby splits the infinitive phrase in two.
[19] Compare this example to ll. 298–300, where the same meaning is expressed in a different order: TVSO. In this context, the prince reveals to his wife the cause for his distress. The Target is encoded by an enclitic pronoun.

(20) **TSOVO** (Sogdian, *Vessantara Jātaka* ll. 887–889) [Benveniste 1946: 56]

mana=ti	vay-a	əxu	ptər	xutāw	əku	dandarak
me=DS	sir-VOC	ART.NOM.SG.M	father	king	to	PN

yar-u	sār	əzdyu	framāt=δār-t	xərti
mountain-ACC.SG.M	towards	exile	order.PP=PRF-3SG	walk.INF

'Sir, the king father ordered me to go into exile to Mount Dandarak.'

In ex. (21), Mandrī expresses her wishes concerning two men to the Supreme God. The initial placement makes clear who is concerned: 'Concerning my husband, I wish ... and concerning that Brahmin who took the children, I wish ...'.

(21) **TSOV** (Sogdian, *Vessantara Jātaka* ll. 1224–1225) [Benveniste 1946: 73]

əti=**mī**	əwēn	wīr-ē	suδāšan	əxu
that=PC.1SG	ART.DAT.SG.M	husband-OBL.SG	PN	ART.NOM.SG.M

ptər	šivī	xutāw	əxšnām	ϑvār-t
father	PN	king	absolution	give.PRS-3SG

'that the king father Šivī gives absolution to my husband Suδāšan'

In ex. (22)–(25), the verb is placed in the initial position. While this can also be a stylistic feature (cf. Section 2.9), this order seems to appear especially when a new scene or narrative thread begins, i.e., a new topic. In the Middle Persian Zoroastrian text *Mēnōg ī xrad*, this order begins nearly every new chapter (62×, ex. 22).

(22) **VSTO** (Middle Persian, *Mēnōg ī xrad* §2.2) [TITUS][20]

pursīd	dānāg	ō	**mēnōg**	ī	**xrad**	kū ...
ask.PST	wise	to	spirit	EZ	wisdom	that

'The wise man asked the Spirit of Wisdom: [direct speech]?'

In ex. (23), the place of narration shifts from the king's palace to the prince's home, which is explicated by this sentence.

(23) **VST** (Sogdian, *Vessantara Jātaka* ll. 278–279) [Benveniste 1946: 19]

rəti	nukər	wītar	xu	wispəšē
and	now	depart.IPRF.3SG	ART.NOM.SG.M	prince

əku	xēpϑ	šēkn	sār
to	own	palace	towards

'And now the prince left for his own palace.'

---

[20] http://titus.uni-frankfurt.de/texte/etcs/iran/miran/mpers/mx/mx.htm?mx002.htm#MX_ii (accessed May 06, 2020).

Ex. (24) closes the scene in the city of Šivagōš. The following sentence opens the new narrative thread of Suδāšan's journey into exile.

(24) **VST** (Sogdian, *Vessantara Jātaka* ll. 396–399)   [Benveniste 1946: 25]
rəti   nukər   zīwart-ant   əxu   šivī   xutāw=ti
and   now   return.IPRF-3PL   ART.NOM.SG.M   PN   king=and
xa   xutēn=ti   xa   inškatē   **əku**   **šivāgōš**
ART.NOM.SG.F   queen=and   ART.NOM.SG.F   harem   to   PN
kanϑ sār   rāyrāyān
city   towards   crying
'And now (they) returned, King Šivī and the queen and the harem, to the city of Šivagōš, crying …'

Ex. (25) is slightly more complicated. This is not a new scene, since the king's decision concerning his delinquent son is being proclaimed. This proclamation likely answers the question regarding the action that is to be taken. As such, the word that expresses the action comes first.[21]

(25) **VST** (Sogdian, *Vessantara Jātaka* ll. 264–265)   [Benveniste 1946: 19]
šaw-a=ti   təyu   **əku**   **dandarak**
go-SBJV.2SG=DS   you.NOM.SG   to   PN
yar-u   əzdyu
mountain-ACC.SG.M   exile
'You shall go into exile to Mount Dandarak!' cf. German 'Gehen sollst du zum Berg Dandarak ins Exil!'

2.8 Focus strategies

It seems that focus, in this case prominence of information within a sentence, influences the placement of constituents. A word under contrastive focus often seems to take the postverbal position.

In ex. (26), the object probably appears at the end due to its length. The Target's position may indicate that it is under contrastive focus. The speaker just revealed that he will go into exile. With this order, he defines the task that he appoints to his wife.

---

[21] Formally, *šawa* could also be an imperative. Since the subject pronoun is present, I assume that this verb form is a 2SG subjunctive.

(26) **VTO** (Sogdian, *Vessantara Jātaka* ll. 300–302) [Benveniste 1946: 20]
rəti frīyān frītam xutēn-ē əpštāy-ām=əskun
and dear.GEN.PL dearest queen-VOC.SG entrust-IND.1SG=PROG
təwa ō zāk-t=ti ō ēškatē
you.OBL.SG ART.ACC child-PL=and ART.ACC harem
'And to you, o dearest queen, (I) entrust the children and the harem.'

The Middle Persian ex. (27) represents the complaint of a sinful wife to her pious husband, who failed to teach her right from wrong. Two constituents of two symmetrically constructed clauses are set into contrast ('from sin' vs. 'towards virtue'). Due to the ergative construction, the object is indicated by the auxiliary (the agreement marker), while the subject is an oblique enclitic pronoun.

(27) **SV-AGR-T** (Middle Persian, *Wizīdagīhā ī Zādspram* 35 §42)
  [Gignoux and Tafazzoli 1993: 136]
u=t[22] bē <nē> wardēnīd ham az wināh
and=PC.2SG away NEG make_turn.PST COP.PRS.1SG from sin
u=t nē hāxt[23] ham **ō** **kirbag**
and=PC.2SG not guide.PST COP.PRS.1SG to virtue
'And you did not turn me away from sin and you did not guide me towards virtue!'

In ex. (28), the disappointed Brahmin laments that he undertook a long journey in vain and will return home empty-handed. Placing the Target in postverbal position may place emphasis on the adverb *wārē* 'empty(-handed)'.

(28) **VT** (Sogdian, *Vessantara Jātaka* ll. 578–580) [Benveniste 1946: 39]
əzu čann wiϑvīt zāy āyət=im
I from far_away land come.PST=COP.PRS.1SG
par məzēx čīnāk ərti əkšī wārē šaw-ām=kām
on great expectation and now empty go-IND.1SG=FUT
əku **xānā** **sār**
to house towards
'I have come from a land far away with great expectations and now (I) will (have to) go home empty-handed!'

---

[22] Emended from 1SG =*m*, which is a form that would neither make sense nor match parallel passages in other texts (e.g., *Ardā Wirāz nāmag* ch. 68).
[23] This word can also be read *āhixt*, which would mean 'draw, pull'. Both readings yield the same structure.

In ex. (29), a group of people is the subject of the subclause, but only one of them is the subject of the main clause, which follows its verb. The postverbal position may be due to this contrast, and the long object phrase comes last.

(29) **TVSO** (Sogdian, *Vessantara Jātaka* ll. 953–956) [Benveniste 1946: 59]

rəti	čānō	əku	kāz-ē	pərēs-ant	rəti=**šan**	āvər
and	when	to	hut-OBL	arrive-3PL	and=PC.3PL	bring.3SG.INJ

əxu	ṛšī	brāman	məyδō=ti	wēx=ti	wərkər=ti
ART.NOM.SG.M	R̥ši	Brahmin	fruit.ACC=and	root=and	leaf=and

āp	čiwēδ	āδču	čuti	xuti	xur-ē
water	from.DEF	something	what	self	eat-OPT.3SG

'And when (they) arrived at (his) hut, the R̥ši Brahmin brought them fruits and roots and leaves and water from whatever (he) himself would eat.'

The Middle Persian *Ardā Wirāz nāmag* exhibits very few postverbal subjects. When Wirāz narrates that exalted spirits came to meet him in the afterlife, he places the subject in final position. This probably highlights the subject (ex. 30).

(30) **TVS** (Middle Persian, *Ardā Wirāz nāmag* 4 §1) [Gignoux 1984: 47]

pad	ān	ī	fradom	šab	**man**	ō	**padīrag**	be	mad
at	that	EZ	first	night	me	to	towards	PFV	come.PST

srōš	ahlaw	ud	ādur	yazd
PN	righteous	and	fire	god

'In the first night, it came towards me the righteous Srōš and the God Fire.'

The Target in ex. (31) consists of four coordinated noun phrases. The first phrase appears after the verb and before the object, while the remaining phrases are moved to the end. This peculiar ordering suggests that special focus is placed on the Target.

(31) **VTOT** (Sogdian, *Vessantara Jātaka* ll. 1443–1445) [Benveniste 1946: 83]

rəti	var-a	**wēšnu**	**vay-ān**	nəmāčyu=ti
and	bring-IPRF.3SG	those.OBL	god-OBL.PL	homage=and

əwēn	yarčīk-t-ē	čētē-tē=ti	əwī
ART.DAT.M	mountainous-PL-OBL	spirit-OBL.PL=and	ART.DAT

xāxīk-t	čētē=ti	əwī	wanē	čētē
source.ADJ-PL	spirit=and	ART.DAT	tree	spirit

'And (he) paid homage to the gods and to the spirits of the mountains and the spirit of the sources and the spirit of the tree.'

The following ex. (32) with a postverbal subject appears in a passage where Mandrī asks the prince three times before he finally replies. The varying word ordering probably underlines the contrast of Mandrī asking and the prince refusing to reply. The context is given schematically below:

TSVO	Mandrī asked **him** [PC] + direct speech as object.
TSV +	The prince was impassive **towards her** [PC]
TOV	and replied nothing **to her** [PC].
TV**S**	Mandrī asked **him** [PC] a second time (ex. 32).
SV	The prince did not reply.
V**S**	Mandrī asked a third time.
TVO	(He) said **to her** [PC] + direct speech as object.

The Targets are encoded by enclitic pronouns, which is why they are realised in initial position. Complement clauses as objects appear at the end. Postverbal subjects appear twice when Mandrī insists on an answer.

(32) **TVS** – O omitted (Sogdian, *Vessantara Jātaka* ll. 290–291)

[Benveniste 1946: 20]

rəti=**šī**=mas	δvityu	əps-a	xa	wəδwa	mandrī
and=PC.3SG=again	second	ask-IPRF.3SG	ART.NOM.SG.F	wife	PN

'And (his) wife Mandrī asked him again a second (time).'

2.9 Style, genre, and calques

In translation contexts, scribes tend to follow the syntax of the *vorlage*. This is illustrated by ex. (33) and (34). In ex. (33), the translator imitates the Avestan syntax by doubling the subject. First, an enclitic pronoun appears to satisfy the rules of Middle Persian grammar, then the coreferential noun appears in the same position as in Avestan. This structure could have been adopted as 'good style' in Zoroastrian texts.

(33) translation Avestan > Middle Persian
(Yasna 9.1 e-f, Manuscript J2, p. 163 l. 15 – p. 164 l. 2)

Avestan	ā	=dim	pərəsaṯ	(z)ara(ϑ)[u](š)trō
	and	=PC.3SG.ACC	asked	Zarathustra
Middle Persian	u=š	az ōy	pursīd	Zardūšt
	and=PC.3SG	from him	asked	Zarathustra

'and Zarathustra asked him'

The Christian Sogdian texts predominantly show verb-first orders, which is a peculiarity of this text corpus. Ex. (34) illustrates how the Sogdian version closely follows the Syriac *vorlage* without violating Sogdian grammar. Few adaptations suggest that imitating the Syriac word order did not result in ungrammatical Sogdian syntax or it may have been adapted as well.[24]

(34) translation Syriac > Sogdian   [Bible, Matthew ch. 21, §37][25]
   Syriac   šaddar   ləwāṭ=hon   la=br=ēh
            ܫܕܪ   ܠܘܬܗܘܢ   ܠܒܪܗ
            send.3SG.PRF toward=PC.3PL.M   to=son=PC.3SG.M
   Sogdian fšamδār-t   ku   wēšan-t   sā   xēpϑ   zātī
            send.PRF-3SG to   them-PL   towards own   son
            'he sent his son to them'   [Barbati 2016: 126, verso l.5]

Similar to translations of Christian texts, the style of Manichaean hymns may have developed in a Syriac-Aramaic context. Iranic Manichaean hymns may either be direct translations of a Syriac-Aramaic *vorlage* or imitate the style of such translations. The following Parthian Manichaean examples all show the verb in initial position. The subject usually follows the verb (95%).

(35) **VSOT** (Parthian, M33 II R.ii.5–7)
        [Durkin-Meisterernst 2006: 50, ll.527–529, and p.51 §d]
   dā(d=i)š   grīw   ō   [duš]men-īn   pad   band
   give.PST=PC.3SG soul to   enemy-PL   in   fetter
   hō   hamag   šahrδārīft
   that   all   kingdom
   'He gave his soul to the enemies, in bondage the whole realm.'
                                         [Durkin-Meisterernst 2006: 51]

(36) **VTS** (Parthian, M4b I V.3–5)   [cf. Boyce 1975: 161 text cv §9]
   āyad   ō   **man**   framanyūg   kū   čīd
   come.PST   to   me   hope   where   always
   būd   hēm   abestaft
   be.PST   COP.PRS.1SG being_under_pressure
   'Hope came to me whenever I felt oppressed.'

---

24 In contrast to the Syriac *vorlage*, the Sogdian Target phrase exhibits a postposition, the object does not exhibit a preposition, and 'own' corresponds to the possessive clitic.
25 The Sogdian version follows Barbati's edition (2016: 125 line 165). Simplified Syriac glossing after the Peshitta Bible, available at http://dukhrana.com/peshitta/index.php (A= status absolutus, C = status constructus, E = status emphaticus). The words in Syriac script would be aligned right to left.

Ex. (37)–(38) display instances where the subject does not follow the verb directly. In ex. (37), a personal pronoun that encodes old information goes between the verb and a subject that encodes new information.

(37) **VTSO** (Parthian, *Angad Rōšnān* 8 §13) [Boyce 1954: 170]
(u)d (w)āxt **(ō) man** *bōžāgar ❖ gyān wēn-ā niδā(ma)g*
and say.PST to me savior soul see-SBJV.2SG husk
(čē wi)zād pad žafr ❖ pad abnās [ud i]spāw
what abandon.PST in depth in destruction and terror
'And the savior said to me: O soul, (you) shall see the husk, which is abandoned in the depth, in destruction and terror!'

In ex. (38), the relative clause of the object is dislocated to the end. It is peculiar that verb and object precede the subject.[26]

(38) **VOST** (Parthian, M5569 V.9–12) [cf. Boyce 1975: 48 text p §4]
ud aĭ paš frēštag parnivrān dād im wigāhīft
and from behind apostle blessed_death give.PST this testimony
uzzi ammōĭag ō **hamag dēn** *čē=š* **dīd andar ispīr**
PN teacher to all religion REL=PC.3SG see.PST in sphere
'And after the blessed death of the apostle, the Teacher Uzzi gave this testimony to the whole church, which he saw in the sphere.'[27]

3 Summary and outlook

Several parameters of morphosyntax, information structure and style determine the placement of Targets. However, without a quantitative analysis that considers the placement of all kind of constituents, it is impossible to decide which, if any, of these placement rules are specific to Targets.

Jügel (2015: 269) notes that ca. 40% of all clauses in the Bactrian corpus, ca. 51% in the Parthian corpus, and ca. 25% in the Middle Persian corpus do not end with the finite verb. Table 8 shows that postverbal Targets form only a small

---

26 In principle, an ergative construction can be interpreted as a passive as well. Then the subject would follow the passive verb. However, in a passive construction, the agent is usually introduced by the preposition 'from' (Parthian *aĭ*, Persian *az*, Bactrian *ac*). Thus, it is unlikely to explain this word order by passive diatheses (verb – subject – agentive phrase – Target).
27 Boyce (1975: 48) offers a more interpretative translation: "(about that) which was seen by him among the soldiery" following Andreas and Henning (1934: 862 fn.2).

part of postverbal constituents and that, if removed together with clausal objects, many other elements appear in postverbal position: 25.5% in Bactrian, 40.5% in Parthian, and 14% in Middle Persian.

Table 8: Percentage of postverbal elements in the corpus of Jügel (2015).

	Postverbal total	Postverbal object clauses	Postverbal Targets
Bactrian	40%	13%	1,5%
Parthian	51%	7%	3,5%
MP	25%	10%	1%

The question of theoretical implications of word order variation is beyond the scope of this paper. However, the multitude of factors that seem to influence word order suggests that word order in Middle Iranic reflects the complex interaction of various grammatical categories. This would imply that speakers of the four Middle Iranic languages decide the status of a participant (e.g., topical or focused) and its encoding (e.g., whether old or new information), which then partly determines the order in which participants appear. The remaining parts of a clause may be placed on the base of semantic or cognitive parameters like iconicity or distance (for the latter cf. Payne and Wasow, both this volume). If so, then word order is the symptom of the underlying information structure that results from various grammatical categories such as topic and focus. Word order can be a diagnostic tool to identify them, but it is not a grammatical category that expresses them.

In terms of whether postverbal placement of Targets is an areal feature, we can note that even if nearly all word order patterns have been attested, most Targets appear before the verb. The preference for Target placement before the verb in Middle Iranic contrasts with the strong tendency for the immediate postverbal position found in several modern Iranic languages. The many postverbal Targets in Parthian and Sogdian translations of Semitic texts support Haig's assumption (2017: 410) that contact with Semitic was important for the development of the "Mesopotamian OVG [i.e., OVT, TJ] word order". However, OVT word order is in itself neither new nor areally confined among Iranic languages. Thus, instead of an 'areal epicenter', it may rather be more appropriate to speak of the prevalence of postverbal Targets in the greater Mesopotamian area, where language contact possibly reinforced a development inherent to Iranic languages.

The tendency for postverbal Targets may have begun at an early stage. The following example is one of the few where the Target appears in the postverbal position for no other reason than being a Target. The subject of the second clause is under contrastive focus as the referent does not seize power but leaves and cedes power to someone else. Focus is achieved by use of the reflexive pronoun *xwad*. The Target of the motion verb refers to the abode of the subject referent and is known information. Since there is no apparent reason for dislocation of the Target, this example may be one of the first that displays the preference of Targets (notably Goals) to follow the verb in a text where Semitic influence can be excluded.

(39) **SVT – OTV** (Middle Persian, *Bundahišn* 33 §11) [Anklesaria 1956: 274]
andar   ham   hazārag   kayhusraw frāsyāb   ōzad
in   same   millennium   PN   PN   kill.PST
'In the same millennium, Kayhusraw killed Frāsyāb,
*xwad šud   ō   **kangdiz** ud   xwadāyīh   ō   **luhrāsp** dād*
self   go.PST   to   PN   and   sovereignty to   PN   give.PST
(he) himself went to Kangdiz and gave the sovereignty to Luhrāsp.'

Abbreviations

1, 2, 3	1st, 2nd, 3rd person
ACC	accusative
ADD	additive marker
AGR	agreement marker
ART	article
COP	copula
DAT	dative
DEF	definite
EZ	*ezāfe* particle (binding attributes to their antecedent)
FUT	future
GEN	genitive
IND	indicative
INF	infinitive
INJ	injunctive
IPRF	imperfect
M	masculine
NOM	nominative
OBL	oblique
OPT	optative
PC	person-marking clitic
PFV	perfective

PL	plural
PN	personal/place name
PP	past participle
PRF	perfect
PROG	progressive
PST	past
REL	relativizer
SBJV	subjunctive
SG	singular
SOTV	subject – object – Target – verb
VOC	vocative

References

Andreas, Friedrich Carl & Walter Bruno Henning. 1934. Mitteliranische Manichaica aus Chinesisch-Turkestan III. *Sitzungsberichte der (königlich-)preußischen Akademie der Wissenschaften, philologisch-historische Klasse* 27. 848–912.

Anklesaria, Behramgore Tahmuras. 1956. *Zand-Ākāsīh – Iranian or Greater Bundahišn*. Bombay: s. n.

Asadpour, Hiwa. this volume. Word order in Mukri Kurdish – the case of incorporated targets.

Barbati, Chiara. 2016. *The Christian Sogdian Gospel Lectionary E5 in Context*. Vienna: ÖAW.

Benveniste, Emile. 1946. *Vessantara Jātaka – Texte sogdien édité, traduit et commenté*. Paris: Librairie orientaliste Paul Geuthner.

Boyce, Mary. 1954. *The Manichaean Hymn-Cycles in Parthian*. London: Geoffrey Cumberlege/Oxford University Press.

Boyce, Mary. 1975. A reader in Manichaean Middle Persian and Parthian. *Acta Iranica* 9.

Čunakova, Oljga Michajlovna [Чунакова, Ольга Михайловна]. 1987. *Kniga dejanij Ardašira syna Papaka* [*The book of deeds of Ardashir, son of Papak*]. Moskau: NAUK.

Dryer, Matthew S. 2013. Order of subject, object and verb. In Matthew S. Dryer & Martin Haspelmath (eds.), *The World Atlas of Language Structures Online*. Leipzig: Max Planck Institute for Evolutionary Anthropology. Available online at http://wals.info/chapter/81 [last access October 17, 2020].

Durkin-Meisterernst, Desmond. 2006. *The Hymns to the Living Soul*. Turnhout: Brepols.

Gignoux, Philippe. 1984. *Le livre d'Ardā Vīrāz – translitération, transcription et traduction du texte pehlevi*. Paris: Ed. Recherche sur les Civilisations.

Gignoux, Philippe & Ahmad Tafazzoli. 1993. *Anthologie de Zādspram – Edition critique du texte pehlevi*. Paris: Association pour l'Avancement des Études Iraniennes.

Haig, Geoffrey. 2017. Western Asia: East Anatolia as a transition zone. In Raymond Hickey (ed.), *The Cambridge handbook of areal linguistics*, 396–423. Cambridge: Cambridge University Press.

Haig, Geoffrey & Geoffrey Khan. 2019. Introduction. In Geoffrey Haig & Geoffrey Khan (eds.), *The languages and linguistics of Western Asia – An areal perspective*, 1–29. Berlin & Boston: De Gruyter Mouton.

Jügel, Thomas. 2015. *Die Entwicklung der Ergativkonstruktion im Alt- und Mitteliranischen – Eine korpusbasierte Untersuchung zu Kasus, Kongruenz und Satzbau*. Wiesbaden: Harrassowitz.
Jügel, Thomas. 2016. Enclitic pronouns in Middle Persian and the Placeholder Construction. *Quarterly Journal of Language and Inscription* 1(1). 41–63.
Kotwal, Firoze M. 1969. *The supplementary texts to the Šāyest nē-šāyest*. Copenhagen: Munksgard
Lambrecht, Knud. 1994. *Information structure and sentence form*. Cambridge: Cambridge University Press.
MacKenzie, David Neil. 1993. *Mēnōg ī Xrad*. Online available at http://titus.uni-frankfurt.de/texte/etcs/iran/ miran/mpers/mx/mx.htm [last access July 28, 2020].
Manuscript J2, available online at http://titus.uni-frankfurt.de/texte/iranica/avestica/j2/j2/j2.htm [last access April 9, 2020].
ʿOryān, Saʿīd (2001 [1382 h.š.]. *Matnhā-ye pahlavī*. Tehrān: Sāzemān-e mīrāṣ-e farhangī-ye kešvar.
Pane, Doris L. this volume. Proximal to distal: Information flow and order in Maa.
Samvelian, Pollet. 2012. *Grammaire des prédicats complexes – les constructions nom-verbe*. Cachan: Lavoisier.
Schmitt, Rüdiger. 1989. Iranische Sprachen: Begriff und Name. In Rüdiger Schmitt (ed.), *Compendium Linguarum Iranicarum*, 1–3. Wiesbaden: Reichert.
Sims-Williams, Nicholas. 2007. *Bactrian Documents II – from Northern Afghanistan: Letters and Buddhist Texts*. London: The Nour Foundation.
Sims-Williams, Nicholas. 2012. *Bactrian Documents I – from Northern Afghanistan: Legal and Economic Documents* (revised edn.). London: The Nour Foundation.
Skjærvø, Prods Oktor. 2009. Old Iranian. In Gernot Windfuhr (ed.), *The Iranian languages*, 43–195. London: Routledge.
Sundermann, Werner. 1981. *Mitteliranische manichäische Texte kirchengeschichtlichen Inhalts*. Berlin: Akademie Verlag.
Wasow, Thomas. this volume. Factors influencing word ordering.

Hiwa Asadpour
Word order in Mukri Kurdish – the case of incorporated Targets

Abstract: In Mukri Kurdish, Targets (e.g., Goals of MOTION verbs, Recipients of GIVE verbs and Addressees of SAY verbs) can appear inside the verbal complex, which is referred to as the 'incorporated position'. The characteristics of this special word order are analyzed based on fieldwork and published data of narrative free speech as well as experimental crowdsourced data (since 2016). Through corpus analysis, I examine morphosyntactic, semantic, discourse-pragmatic and cognitive factors that trigger word ordering. The results show that the incorporated position in Mukri is different from other positions in terms of syntactic dependency length, animacy and adjacency of verb + non-verbal elements. Nevertheless, the incorporated position can be seen as a variant of the preverbal position.

Keywords: incorporation; multi-word verbs; Mukri Kurdish; Target; word order

1 Introduction

This article discusses the morphosyntactic and semantic features of incorporated Targets in Mukri. 'Target' is an umbrella term for the semantic roles of constituents in constructions with two or three-place verbs, including Destinations of MOTION and CAUSED-MOTION verbs, e.g., *çûn* 'to go', *dânân* 'to put'; Recipients of GIVE verbs, e.g., *dân* 'to give'; Addressees of SAY verbs e.g., *gûtin* 'to say'. All of these constituents are usually flagged with adpositions[1], which express

---

[1] The term adposition is used as an umbrella term to cover pre-, post- and circumpositions including their absolute and clitic variants. An 'absolute adposition' is the form of an adposition that appears if the complement is a bound morpheme, e.g., an enclitic pronoun, *be to* vs. *pê=t* both 'to you'. 'Primary adposition' refers to the simple adpositions (such as *be*) and 'compound adposition' to the combination of a primary adposition with another element, typically a noun (such as *be ser*). The term flagging is also used for adpositional/case marking and the verb *to flag* means to mark an element with a case or an adposition. Other alternative is marking, but this term is used for marking definiteness in my study.

Hiwa Asadpour, Department of Linguistics, Institute of English and American Studies (IEAS), Goethe University Frankfurt, Germany/Department of Languages and Information Sciences, University of Tokyo JSPS International Research Fellow, Japan, E-mail: asadpour@lingua.uni-frankfurt.de

https://doi.org/10.1515/9783110790368-004

direction towards an endpoint such as *bo* 'to, for', *be* 'to', *le* 'at' and *de* 'at, into' in GIVE, MOTION[2], SAY, and CAUSED-MOTION verbs; *=e* 'to' in all of the verb types; and their absolute variants such as *bo, pê, lê, tê* and *=ê*. Since Beneficiaries as well as Destinations of CHANGE-OF-STATE, SHOW and LOOK verbs[3] are flagged with the same adpositions, they are included in the survey as well. Targets in the incorporated position can also be flagged by an *ezâfe*[4] (see Section 2.1) or be left bare in the postverbal position, i.e., no flagging (see Section 3). Full flagging refers to the full forms of adpositions (such as *bo*) and reduced flagging refers to the cliticized forms (such as *=e*).

In general, Targets can precede or follow the verb with a strong preference for the postverbal position in Mukri. If a verb consists of more than one word (called multi-word verb (MWV) in this study (cf. Section 2)), Targets can be placed within the predicate construction. This means that they follow the first word of the MWV and precede the second, which is the finite verb, i.e., they are incorporated into the MWV (cf. ex. 1 and Section 2). This position has not received special attention in previous studies.

(1) Incorporated Target: *ḧerz kirdin* (lit.) 'to tell' [OD]

S	NVE[5]	P	T	V
emin	ḧerz	**be**	**to**	de-ke-m
I	request	to	you.SG	IPFV-do.PRS-1SG

'I am telling you.'

The main questions about incorporated Targets are: Can the incorporated position be considered a variant of the preverbal position, because the Target appears before the finite verb, or is it an independent position, because it appears inside the verbal complex? In what kind of functions and contexts do incorporated Targets differ from pre- or postverbal Targets?

---

2 In the majority of Mukri varieties, *bo* is typically used for MOTION verbs instead of *be*. The latter form is more typical in western areas of Mukriyan (cf. Asadpour 2021 for the dialect classification of Mukri).
3 These include verbs such as *bûn* 'become', *nwândin* 'to show', or *çâw lê kirdin* 'to look at'.
4 In Iranian linguistics, the morphological marker that links attributes to the head noun is known as '*ezâfe* marker' (EZ) (cf. Haig 2011: 363–392 for *ezâfe* in Kurdish, with further references).
5 NVE refers to the non-verbal element of the MWV construction (see Section 2).

1.1 Mukri

Mukri is a Kurdish variety of the Iranian language family spoken in the province of Western Azerbaijan in the northwestern part of Iran (Map 1).

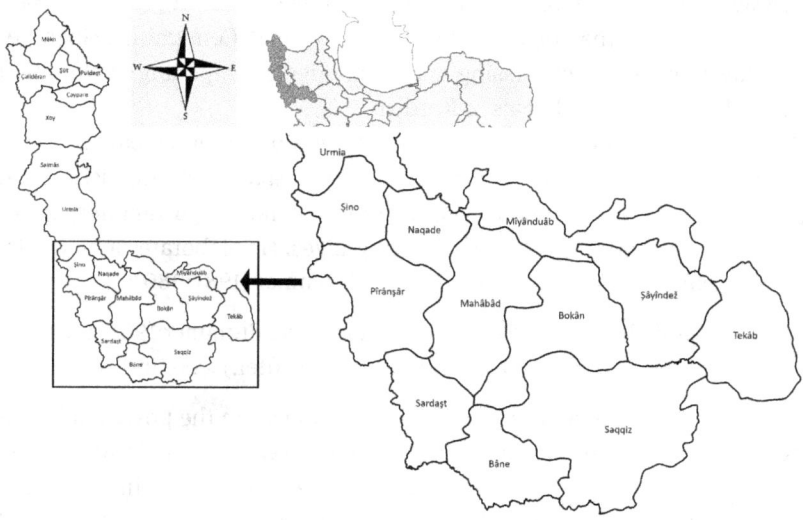

Map 1: Mukriyan region in southern areas of Western Azerbaijan (©2022 Hiwa Asadpour).

In addition to Mukri and Persian, which is the official language and lingua franca of Western Azerbaijan and of Iran as a whole, several other languages have been in contact for centuries in this province: Indo-European (Northeastern Kurdish varieties and Armenian), Semitic (Northeastern Neo-Aramaic), Turkic (Azeri Turkic, also called Azerbaijani).

1.2 Brief overview of Target word order in the literature

One of the first studies that discusses the word order of sentences with constituents such as goals, recipients, and addressees in Iranian languages is Frommer (1981). He explores informal and literary Persian and concludes that postverbal goal arguments are typical for spoken and informal language. Haig (2015, 2017, to appear) investigates Kurdish and other Western Iranian languages and shows that goals, recipients, and "final state arguments" are predominantly in the postverbal position in Kurdish. Haig suggests that postverbal goals could be a universal phenomenon, motivated by iconicity: "all other things being equal, semantic endpoints may be the preferred clausal endpoints" (Haig to appear: Section 6.2, see also Haig 2015 for discussion of other OV languages). Historically, postverbal goals do appear sporadically in Iranian (see Jügel, this volume), but this feature may have been amplified due to contact influence with Aramaic and possibly other VO languages. With regard to theoretical approaches, Haig (to appear) concludes that neither the Final-over-Final Constraint (Biberauer 2017) nor Hawkins' (2008) processing approach offer a compelling explanation for the ubiquity of postverbal goals in Kurdish.

Stilo (2018) gives an overview of the distribution of the preverbal and postverbal placement of "peripheral arguments" in the Araxes-Iran Linguistic Area. His sample comprises 29 languages from different language families, 26 of which are predominantly object-verb (OV) languages. He elaborates a hierarchy for postverbal elements in his sample based on Frommer (1981: 180):

> Postposability hierarchy (Stilo 2018: 5 modifying Frommer 1981: 180)
> Goal > Recipient > Beneficiary > Addressee > DO (def.) > DO (indf.)

The hierarchy shows that goals are most likely to appear in the postverbal position, then recipients, then beneficiaries, then addressees, then a definite direct object 'DO (def.)', and least likely an indefinite direct object 'DO (indf.)'. Stilo concludes that not all 'oblique arguments' are likely to occur postverbally, and the likelihood of postverbality of these elements depends on the geographical location of the languages under investigation. Several other studies concerning various Iranian languages mainly focus on pre- and postverbal positions of goals, among other elements (Korn, this volume; Jahani 2018; Gündoğdu 2018: 149–179). Faghiri and Samvelian (2021: 117) mention the inseparability of complex predicates and consider this issue as a word order variation phenomenon. However, their topic is not related to the incorporated position in this study. This previous research led me to consider not only conventionally syntactic and semantic, but also discourse-pragmatic and cognitive factors in the study of Target word ordering in this study.

1.3 Data collection

I collected the examples for incorporated Targets in Mukri from two sources: (a) Öpengin's (2016) Mukri corpus (ÖM), and (b) from my own multilingual corpus of Target Order in Northwestern Iran (TONI corpus), which includes data from Mukri Kurdish, NEK varieties, Armenian, Azeri, and partly Christian NENA. The corpora represent narrative free speech, such as tales and memories.

In addition, I field-tested Target examples in Figure 1–4 with experimental methods by crowdsourcing grammaticality judgments.[6] This included face-to-face interviews as well as interviews conducted on social media using tools such as Telegram, WhatsApp and Facebook (since 2016). The selected users answered various questions, via text or voice, relating to a special type of word order. Their answers included dialectal phrases together with their personal assessments of a specific Target word ordering. Moreover, the informants were also asked to evaluate their own interpretations by assessing those of other community members. The questions for the crowdsourced data took various forms, for example by presenting a list of different options for the word order of a single sentence. The participants could answer each of my questions in one of three ways: (a) text message, (b) voice message, and (c) icon/emoji messages – using a single emoji of surprise, question, excitement, etc.

For referencing my unpublished fieldwork, I use '[OD]' for crowd-sourced data and the initials of the informants' name plus text paragraph number for free-speech data.[7]

2 Incorporated position of Targets

The incorporated position can be realized in two ways: (a) the Target attaches to the verbal complex as a bound morpheme (not explored in this study), and (b) it is placed within a multi-word verb (MWV) in the form of a noun or an orthotone pronoun. The term MWV covers combinations that form a semantic unit of a verb and a non-verbal element (NVE), viz. nouns and adpositional phrases.[8] These combinations show placement constraints with respect to the non-verbal element. If it is a noun, the incorporated Target can be flagged by either an *ezâfe* or an adposition (Section 2.1). Three types of non-verbal elements occur in incorporated constructions: (a) 'nominal elements realized as subject', (b) 'nominal elements realized as object' (Section 2.1), and (c) 'adpositional elements' (Section 2.2).

---

[6] The practice of obtaining information or input into a task or project by enlisting the services of a large number of people, typically via the Internet and social media tools.
[7] For consistency between different corpora used in this study, the transcription, glossing and translation of the sentences in Öpengin (2016) have been slightly modified.
[8] Previous literature discusses MWV of the type 'noun + verb' under the term 'complex predicate' (cf. Haig 2002; Megerdoomian 2012; Samvelian 2012; Gündoğdu 2018: 111–147 for further references; and Mithun 1984 for a survey of typological classification on noun incorporation).

2.1 Nominal element + verb

MWVs with a nominal element are by far the most common type.

swâr bûn 'to mount'

(2) Fixed construction: *swâr bûn* 'to mount' [OD: FM.100a]

NVE	=EZ	T	V	
xerîk=e	swâr(=î	ser)=im	bê	be zor-î
PROG=COP.PRS.3SG	rider=EZ	head=1SG:POSS	be.SBJV.PRS.3SG	with force-OBL

'He is going to force himself on me.'
(lit. '(he) is going to mount on (become a rider of) my head forcefully).'[9]

In (2), the Target *serim* 'my head' is linked by an *ezâfe* to the nominal element *swâr*. Thus, the nominal element is separated by an *ezâfe* and the nominal Target *serim* from the verb. If the Target is the enclitic pronoun *=im* 'me', it attaches to the nominal element directly. The MWV *swâr bûn* 'to mount' forms a semantic unit and differs from combinations such as *mâmostâ bûn* 'to be(come) a teacher', because the semantics of the latter are purely compositional (this is called a 'compositional construction'). In the fixed construction, i.e., ex. (2), the nominal element cannot be modified, while in a compositional construction the noun (here: *mâmostâ*) can be modified because it represents a syntactic unit independent from the verb.

wiłâm dânewe 'to give an answer'

Another example is the MWV *wiłâm dânewe* 'to give an answer'. In this MWV structure, a free morpheme, i.e., *Hemze* (PN), can be linked to the nominal element by an *ezâfe* (3a) or by a preposition (3b). In both cases, the nominal Target is also marked for case.

(3) a. Ezâfe flagging: *wiłâm dânewe* 'to give an answer' [OD]

NVE	=EZ	T	V	
ege ne=y=twânî	wiłâm=î	Hemze-y	bi-dâ-t=ewe	
if NEG=3SG=can.PST	answer=EZ	Hemze-OBL	SBJV-give.PRS-3SG=POSTV	

'If (he) couldn't give an answer to Hemze...'

b. Prepositional flagging: *wiłâm dânewe* 'to give an answer' [OD]

NVE	P	T	V	
ege ne=y=twânî	wiłâm	be	Hemze-y	bi-dâ-t=ewe
if NEG=3SG=can.PST	answer	to	Hemze-OBL	SBJV-give.PRS-3SG=POSTV

---

[9] If the complement of *ser* is a female person, the verb can also be understood as 'to harass'. For example, *ew dûkândâre zor be çâw=û ruwewe xerîke swârî sermân debêt* 'This shopkeeper is really harassing us.'

As these constructions (3a–b) do not differ semantically, the *ezâfe* can be considered a flagging variant of the adposition.

In examples (3a–b) the nominal element *wiłâm* can be quantified, determined or modified, for instance with plural marking, e.g., *çend wiłâm-ekân=î Hemze-y* 'several answers to Hemze'. It displays the syntactic properties of a direct object, which excludes interpreting *wiłâm dânewe* as a fixed expression. Therefore, examples (3a–b) can be considered compositional constructions.

However, *wiłâm dânewe* can also be used in a construction where it is not possible to modify the nominal element (4).[10]

(4) Fixed construction: *wiłâm dânewe* 'to answer' [OD]

	O		P	T	NVE	V	
ew	sê	pirsyâr=e-y	**be**	**min**	wiłâm	de=yewe	
this	three	question=DEM-OBL	to	me	answer	give.IMP.2SG=POSTV	

'Answer these three questions to me!'

In (4), *pirsyâr* fills the direct object position and *wiłâm* appears petrified in a fixed construction with the verb. In such fixed constructions, the incorporated position is blocked for the Target, i.e., the Target is either in preverbal or postverbal position. As a fixed construction, *wiłâm dânewe* can be used in a figurative sense, i.e., 'to react', as in example (5) below:

(5) Fixed construction: *wiłâm dânewe* 'to react' [OD]

		T		NVE	V	
ew	**kar=e=y**	**Hemze=y**		wiłâm	dâ=yewe	
this	action=DEM=EZ	Hemze=PC.3SG		answer	give.PST=POSTV	

'(He) reacted to this action of Hemze.'

If the nominal element is followed by an adjective in an *ezâfe* construction, the adjective does not modify the noun, but the MWV as a whole (ex. 6). In such a construction, the Target may still be in the incorporated position linked by an *ezâfe* and modified by an adjective.

(6) Adverbial modification: *wiłâm dânewe* 'to react' [OD]

NVE	ADV	=EZ	T	V
wiłâm=î	tund=î	**Hemze-y=î**	**rûreş=î**	dâ=y=ewe
answer=EZ	strong=EZ	Hemze-OBL=EZ	shamed=3SG	give.PST=3SG=POSTV

'(He) reacted severely to the shamed Hemze.'

---

10 Cf. Samvelian (2012: 55–87) for a detailed account on the status of nominal preverbs in Persian.

The examples (3)–(6) show that constructions such as *wiłâm dânewe* are ambiguous because they can be considered a bridge between a compositional and fixed MWV.

In some MWVs, the nominal element appears as the morphological subject and the Target as a topic (cf. the literal translation of ex. (7a)).

bon hâtin 'wafting smell'

(7) a.   NVE as subject: *bon hâtin* 'wafting smell' [ÖM: 200, ČQ.27]

	NVE/S	=T	P	V	
heç-kes-êk-î	bon=îş=î		bo	bê	her
any-one-INDF-REL[11]	smell=ADD=3SG		to	come.SBJV.FUT.3SG	definitely

demrê
die.IPFV.PRS.3SG

'Anyone who smells (it) will definitely die!' (lit. 'Anyone whom smell comes to will definitely die!')

In (7a), the nominal element of the MWV, i.e., *bon*, is formally the subject of the relative clause. The enclitic pronoun *=î* is the Target. It is governed by the preposition *bo* and refers to *heçkesêkî*, which is the subject of the main clause. Compare with (7b), which displays the same expression in a main clause.

(7) b.   NVE as subject: *bon hâtin* 'wafting smell' [OD]

(T)	NVE/S	=T	P	V	
(min)	bon=î	roz-ekân=**im**	bo	dê	
I	smell=EZ	rose-DEF.PL=1SG	to	come.IPFV.PRS.3SG	

'I smell the roses (lit. the smell of the roses comes to me).'

Here, too, *bon* is the grammatical subject and triggers agreement with the verb. The Target is the enclitic pronoun *=im*. Since the Target in such expressions is usually topical, there is a strong tendency to double it with an orthotone pronoun in a hanging-topic position, i.e., *min* in (7b). This relation of hanging-topic and enclitic pronoun, which can be described as topic agreement, is similar to verbal agreement in the post-ergative construction of Mukri (cf. Öpengin 2016: 53–55, 82–84 for the agreement function of pronominal clitics in Mukri). In such constructions, the experiencer (i.e., the Target) tends to adopt subject properties (see Jügel and Samvelian 2020: 148 on Persian). There are many other similar examples in the ÖM and TONI corpora, for instance:

---

[11] This is commonly referred to as *ezâfe* as it is commonly identical.

(8) NVE as subject: *xew lêkewtin* 'falling asleep' [ÖM: 258, ČN.217]

| NVE/S =T P V |

hînde hîlâk=û mândû bû xew=î lê kewt
so tired=and exhausted COP.PST.3SG sleep=3SG at fall.PST.3SG

'(He) was so tired and exhausted (that) he fell asleep [*lit.* sleep fell on him].'

2.2 Adpositional element + verb

The second type of MWV consists of an adposition and a verb where the adposition is obligatory for the semantics of the MWV (e.g., *lê=dân* (lit.) 'to give at' meaning 'to beat on, to knock at'). These adpositional MWVs cannot be formally distinguished from compositional constructions of verbs that appear with an adpositional phrase which is not obligatory (e.g., *lê=kutân* 'to beat at'). Both consist of a verb and an adposition and allow the Target to appear in between. The adpositional element and the Target can appear in the postverbal position as well. However, in contrast to *lêkutân*, the adposition of the MWV *lêdân* can neither be deleted (9a vs. 9b) nor cliticized (9c vs. 9d) and must appear adjacent to the verb. Therefore, *(lê)kutân* is classified as a simple verb with an adpositional phrase and *lêdân* as a MWV.

(9) a. Adpositional NVE: *lêdân* (lit.) 'to give at' [OD]

| NVE T V |

le dergâ-ke=yân de-de-m
at door-DEF=3PL IPFV-give.PRS-1SG

'(I) am knocking at their door' (lit. 'I give at the door').

In (9a), the Target is placed between the adposition *le* and the verb *dedem*. Omitting the adposition is unacceptable semantically and grammatically, in contrast to (9b).

(9) b. Verb + preverbal adpositional phrase: *kutân + lê* 'to beat + on' [OD]

| P T V |

(le) dergâ-ke=yân de-kut-im
at door-DEF=3PL IPFV-beat.PRS-1SG

'(I) am beating on their door.'

Semantically (9b) carries only the literal meaning 'I beat on the door', while (9a) includes other actions like calling out to someone to come and open the door.

The Target can occur in the postverbal position, where the adposition can occur in four ways: (a) unchanged (*le*), (b) substituted by the enclitic variant

(=*e*), (c) together with the enclitic variant (=*e le*),[12] and (d) omitted. The preferred form is either to have the preposition *le* or its enclitic variant =*e*.[13]

(9) c. Verb + postverbal adpositional phrase: *kutân* + *lê* 'to beat + on' [OD]

V	(=P)	(P)	T	
de-kut-im(=e)		*(le)*	*dergâ-ke=yân*	
IPFV-beat.PRS-1SG(=on)		(at)	door-DEF=3PL	

'(I) am beating on their door.'

In (9c), the adposition *lê* in *lêkutân* can be omitted or cliticized to the verb without change its meaning or violating grammatical rules, in contrast to *lêdân* (9d).

(9) d. Postverbal adpositional NVE: *lêdân* [lit.] 'to give at' [OD]

V	=NVE	T	
*de-de-m=e		*dergâ-ke=yân*	
IPFV-give.PRS-1SG=at		door-DEF=3PL	

[Intended meaning] '(I) knocked at their door.'

In general, adpositional elements of MWVs have two functions: (a) showing the direction in the event structure and (b) changing the semantics of the simple verb. Therefore, adpositions cannot be omitted.

2.3 Analysis

My data shows that there are different types of MWVs in which Targets can occur in the incorporated position. For MWVs with adpositional elements, the placement of the Target is not flexible. The Target must follow the adposition and the adpositional phrase must precede the finite verb. In MWVs with nominal elements, the placement of the Target is flexible. Thus, the question arises as to what makes a Target appear in incorporated position.

---

12 Similar to compound adpositions where a second element modifies the primary preposition such as *dekutim=e ser Hemzey* 'I am beating on Hemze', it is also possible to have two primary adpositions together as in example (9c) *dekutim=e le dergâyekeyân*. The primary adposition =*e* is the neutral directive preposition and the second primary preposition *le* clarifies how the movement is happening. Doubling the preposition may be uncommon, but it is not ungrammatical.

13 For instance: *le zîndân-ê=y   nâ*   vs.   *nâ=y=e   zîndân-ê.*
at prison-OBL=3SG put.PST   put.PST=3SG=to prison-OBL
'he put him into prison'

Table 1 shows the frequency of NPs and orthotone pronouns in the incorporated position with their flagging.[14]

Table 1: Target flagging in incorporated position.

Target PoS	Pattern	PREP	EZ	Total
NP	N + T + V	9	6	57
	ADP + T + V	42	–	
PRO	N + T + V	1	12	19
	ADP + T + V	6	–	

There are two possible ways of flagging nouns and pronouns, viz. either by an adposition or by an *ezâfe*. In MWV constructions, in which the non-verbal element is a noun, 89% of nominal Targets are flagged by a preposition and 11% by an *ezâfe*. On the contrary, pronominal Targets are mostly flagged by an *ezâfe* (63%). The Fisher's exact test show statistical significance with respect to the occurrences of NPs and pronouns with preposition and *ezâfe* flagging: $p \approx 0.000016$, i.e., $p < 0.01$.

This difference in flagging is motivated by information structure, which seems to be related to marking focus (neutral focus and contrastive focus[15]). Non-focused Targets appear with the lightest possible flagging: for incorporated Targets this means that, if pronominal, they appear as enclitic pronouns; if nominal, they are flagged by an *ezâfe* ($p < 0.01$). If the Target is under focus, pronominal Targets appear in their orthotone form.[16] *Ezâfe* flagging indicates neutral focus, while contrastive focus is achieved either by adpositional flagging or by *ezâfe* flagging in combination with a specific stress and pitch. The same choice of flagging occurs with nominal Targets, but it then indicates neutral focus. Contrastive focus of nominal Targets is indicated by adpositional flagging together with stress and pitch (cf. Table 2).

---

14 The statistics in Table 2 are based on 708 Target sentences from the entire ÖM corpus with the selection of only nominal and pronominal Targets excluding bound morphemes.

15 Neutral focus is non-presupposed information, which is highlighted, and contrastive focus is a subset of the set of contextually or situationally given elements. As such focus is not restricted to represent new information.

16 Due to the existence of enclitic pronouns and the possibility of pro-drop, the use of orthotone pronouns is marked (cf. Korn 2009: 159; Jügel 2017: 557).

Table 2: Relation of flagging and focus marking in incorporated position.

	Non-focused	Neutral focus	Contrastive focus
nominal T	+ *ezâfe*	+ *ezâfe* + stress & pitch or: + adposition	+ adposition + stress & pitch
pronominal T	enclitic form	orthotone form + *ezâfe*	orthotone form + *ezâfe* + stress & pitch or: + adposition

3 Comparing incorporated Targets with pre- and postverbal Targets

This section investigates whether incorporation is an independent position or a special variant of the preverbal position by comparing flagging (Section 3.1), semantic interpretation (Section 3.2), and the distribution of animacy (Section 3.3). It also tests cognitive principles that predict the optimal placement of constituents to check whether they can offer a compelling explanation for the incorporated position (Section 3.4).

3.1 Flagging

There are three types of flagging: full (full form of adposition), reduced (cliticized form of adposition in postverbal position or *ezâfe* in incorporated position), and bare (no adposition). My investigations show that full flagging is available for all positions (ex. 10, 13a–c). Reduced flagging only appears in the postverbal (cliticized adposition) and incorporated positions (*ezâfe*, ex. 11). Bare flagging only appears in the postverbal position (ex.13a).

(10) Preverbal Target [OD]

| P | T | NVE | V |

ege ne=y=twânî **be Hemze-y** wiłâm bi-dât=ewe
if   NEG=3SG=can.PST to Hemze-OBL answer SBJV-give.PRS.3SG=POSTV

'if (he) couldn't answer to Hemze'

(11) Incorporated Target [OD]

| NVE |=EZ T | V |
ege ne=y=twânî wiłâm=î **Hemze-y** bi-dât=ewe
if   NEG=3SG=can.PST answer=EZ Hemze-OBL SBJV-give.PRS.3SG=POSTV

According to examples (10) and (11), a Target in the preverbal position is flagged by a preposition and marked for case, which matches the findings in my corpus. Contrary to preverbal and incorporated positions, Targets in the postverbal position can be flagged in four ways: (1) by an adposition and oblique case, (2) by only an adposition, (3) by only oblique case or (4) left bare (cf. ex. 12).[17]

(12) Postverbal Target [OD]

O | NVE | V | (P) T |
ew sê pirsyâr=e-y wiłâm de=yewe **(be) Hemze(-y)**
this three question=DEM-OBL answer give.IMP.2SG=POSTV (to) Hemze-(OBL)

'Answer these three questions to Hemze.'

Flagging is obligatory in the preverbal and incorporated positions, while postverbal Targets are mostly bare.[18]

3.2 Semantic interpretation

The semantic interpretation of MWVs differs depending on the position of the Target. For instance, the MWV *we řê kewtin* has a literal meaning 'to fall onto the road' as well as a figurative meaning 'to take to the road, to set off'. With the Target in the postverbal position, the construction is always interpreted literally regardless of flagging (ex. 13a). With the Target in the incorporated or preverbal position, the literal or figurative interpretation depends on flagging (ex. 13b and 13c).

---

17 Eastern Mukri dialects tend to lose case marking. In these dialects, Targets can be unmarked for case in preverbal and incorporated position as well (see Asadpour 2021).
18 Cross-linguistically, similar observations can be related to the blockage within the clausal nucleus (preverbal) vs. flexibility within the clausal periphery (postverbal) positions. Sidwell (2020: 82–104) has discussed this phenomenon in connection to Nicobarese (Austroasiatic) languages.

(13) a. Postverbal Target [OD]

	S	V	(P)	T(=P)
ew-ê şew-ê ke	şâh ʕebâs	kewtbû	**(we)**	**ř̌ê(=we)**
that-OBL night-OBL that	king Abbas	fall.PPRF.3SG	to	road=POSTP

'that night when King Abbas had fallen onto the road ...'

b. Preverbal Target flagged by a preposition [OD]

	S	P	T	V
ew-ê şew=ê ke	şâh ʕebâs	**we**	**ř̌ê**	kewt
that-OBL night=OBL that	king Abbas	to	road	fall.PST.3SG

'that night when King Abbas set off...'

In (13a), the Target in the postverbal position can be flagged by a circumposition (*we ř̌êwe*), a simple preposition (*we ř̌ê*) or it can be bare (*ř̌ê*). If the Target appears in the postverbal position, the construction is always interpreted literally. In (13b) the meaning cannot be interpreted literally like in (13a).[19] If the Target is in the preverbal position, the literal interpretation can only be achieved by a circumposition, as in (13c):

(13) c. Preverbal Target flagged by a circumposition [OD]

S	P	T=P	V	
şâh ʕebâs	**we**	**ř̌ê-yêk=ewe**	kewtbû	be zor-î
king Abbas	to	road-INDF=POSTP	fall.PPRF.3SG	with force-OBL

'King Abbas made a bad fall onto the road.'

Thus, the Target *ř̌ê* can be fully flagged in the incorporated, pre- and postverbal positions, but the interpretation changes in the incorporated and preverbal positions depending on the flagging.

3.3 Animacy

If we consider all kinds of Targets (including adverbials and bound morphemes), they are mostly animate in the incorporated position. This is similar to the preverbal position, while inanimate Targets are preferred in postverbal position (see Table 3).

---

[19] The figurative or idiomatic interpretation derives from the Target reading. It is peculiar that the semantic change only took place with the Target in preverbal position (*we rê kewtin*). This could be due to lack of iconicity (I owe the gist of this thought to Thomas Jügel).

Table 3: Target animacy in different positions (based on Öpengin 2016 corpus).[20]

	Preverbal	Incorporated	Postverbal
Animate T	25%	21%	6%
Inanimate T	16%	8%	23%

Based on Fisher's exact test, animacy is a statistically significant factor that determines word order.

Table 4: Statistical significance of the relation between animacy and Target word ordering.

Incorporated	Non-incorporated	Fisher's exact test
21	31 (25+6)	$p \approx 0.015$, i.e., $p < 0.05$ (fairly significant)
8	39 (16+23)	
preverbal	postverbal	
25	6	$p \approx 0.0013$, i.e., $p < 0.01$ (highly significant)
16	23	

3.4 Summary

Table 5 summarizes the features discussed in Sections 3.1–3.3.

Table 5: Properties of Targets in various positions.

	Position		
	preverbal	incorporated	postverbal
FULL FLAGGING	+	+	+
EZÂFE FLAGGING	−	+	−
REDUCED FLAGGING	−	−	+
BARE	−	−	+
LITERAL INTERPRETATION	−/+	−/+	+
ANIMACY	high	high	low

---

[20] The statistics are based on the entire ÖM corpus with 708 Target sentences.

4 Cognitive principles

In terms of the above word order types and in particular incorporated Targets, the question arises whether cognitive principles trigger Target word ordering. Behaghel (1932: 4) observed "daß das geistig eng Zusammengehörige auch eng zusammengestellt wird".[21] This means that syntactic proximity may reflect semantic closeness (cf. Wasow, this volume, for the definition of semantic closeness). This could explain why the Target tends not to appear in the incorporated position if the nominal element and the verb form is a fixed expression. Since the nominal element and the verb constitute one semantic unit, they are usually placed close together.

Behaghel's generalization was transferred to the relation of verbs and objects in that they tend to occur adjacent to one another (cf. Haiman 1983: 782). Just like the combination of object and verb, which constitutes a semantic unit, the Target can enter a fixed relation with the verb as well.

Semantic closeness of Target and verb seems to improve the efficiency of communication between the speaker and the hearer,[22] which is nowadays referred to as the 'processing principle' (cf. Hawkins 1994; Gibson 1998; Wasow, this volume, for details). Two theories of processing principles are widely considered: a) the Minimize Domains (MiD) principle (Hawkins 2004: 107/109/113), formerly known as Early Immediate Constituent (EIC) (Hawkins 1994), and b) the Syntactic Prediction Locality Theory (SPLT) (Gibson 1998: 19).

(a) Hawkins (1994) discusses the interconnection between flagging, verb and position of the constituents. His proposal originated from Greenberg's generalization (1963) that VO languages prefer prepositions and OV languages prefer postpositions (cf. Greenberg's Universal 4, p. 62; Wasow, this volume).

Hawkins (1994: 257) considers four different word orders within a verbal phrase domain including pre- or postpositional phrases, i.e., [P NP] or [NP P].

A: [V [P NP]] B: [[NP P] V] C: [V [NP P]] D: [[P NP] V]

According to Hawkins (1994: 257), type A and B are the most frequent word orders in the world's languages. He considers them to be optimal ordering because the adposition (the head of the phrase) stands next to the verb. This theory predicts that languages with postpositional flagging prefer left-branching (PP

---

21 Translation by Wasow (this volume): "that what belongs close together conceptually also gets placed close together." Behaghel labels this "[d]as oberste Gesetz" ('the highest law') of word order.
22 Cf. Kemp et al. (2018), who discusses the semantic closeness of object and verb.

in preverbal position), while languages with prepositional phrases prefer right-branching (PP in postverbal position).

Mukri predominantly has prepositions. Postpositions only appear in combination with prepositions, i.e., as part of circumpositions, yet the basic word order is OV. Hawkins' classification does not mention circumpositions. They can be analyzed as consisting of a semantically decisive preposition as the head of the circumpositional phrase and a postposition that only modifies the meaning. *Ezâfe* flagging can be considered a variant of type [P NP] (cf. Section 2.1).

If Hawkins' classification serves as a basis for evaluating the distance between head and verb, MWVs pose a problem. Is the finite verb or the non-verbal element the relevant part for the syntactic dependency (see Section 3.2 for discussion)? The adpositional phrase would either precede (before V: N [ADP T] V) or follow (after N: N [ADP T] V). Table 6 summarizes three cases of Target incorporation.

Table 6: Various incorporated word order patterns in the ÖM corpus of pronominal and nominal Targets.

	Incorporated Targets	77		100%
		NP	PRO	
1	[N [ADP T] V] *wiłâm be Hemzey bidâ (wiłâm dân)* 'he gives an answer to Hemze'	13	1	18%
2	[[adp T] V] *le dergâkey meşit dâwe? (lêdân)* 'Have you also knocked at our door?'	36	10	60%
3	[N[=EZ T] V] *herz=î to bikem (herz kirdin)* 'let me tell you'	8	9	22%

(b) The SPLT principle considers the dependency distance between all elements of a sentence. The words[23] between each head and its complements are counted and summed. The arrows in the figures below display the dependencies between the elements, and the number on top of the arrows indicates the distance. The dependency sum determines the overall dependency length ('8' in Figure 1).

---

[23] In this principle, the term 'word' is not defined. It is unclear if bound morphemes are considered or not, which may influence the results of this analysis. For instance, whether to count *ezâfe* as a word or not remains controversial (cf. Bögel and Butt 2012, with further references).

The assumption is that a smaller sum is 'better' than a larger one. This implies that elements which belong together tend to appear in close proximity to one another, which corresponds to Behaghel's law of semantic closeness.

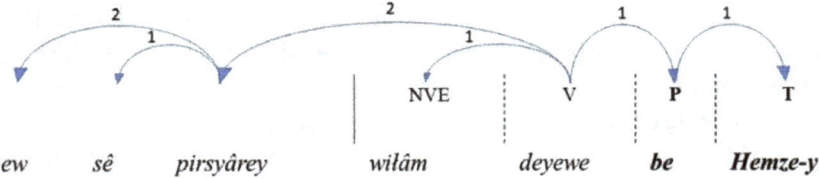

Figure 1: Postverbal Target: total dependency length = 8, IC = 54%.

The sentence in Figure 1 means 'if you answer these three questions to Hemze' (for glossing see ex. 12). Figure 2, Figure 3, and Figure 5 are all variations of this sentence with the Target, i.e., *be Hemzey*, in different places.

According to the SPLT principle, the word order in Figure 1, with a dependency length of 8, is more optimal than that in Figure 2, with a length of 12. Figure 1 also adheres to the EIC/MiD principle, i.e., adjacency of the verb to the adpositional head, in contrast to Figure 2.

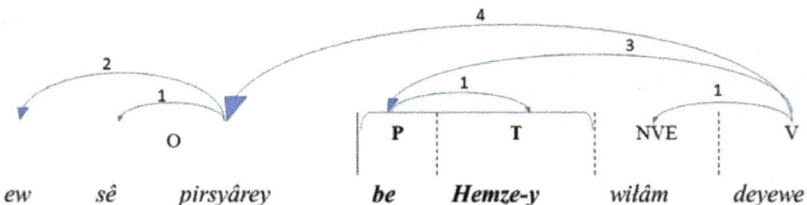

Figure 1: Preverbal Target: total dependency length = 12, IC = 52%.

Similar to Figure 2, Figure 3 fails to fulfill the predictions of the SPLT and EIC/MiD principles as well. The example represents a suboptimal order because shifting the Target to the initial position increases the total dependency length. However, despite the pragmatic marking and all principles considering it suboptimal, all respondents confirm that Figure 3 is an accepted and natural word ordering.

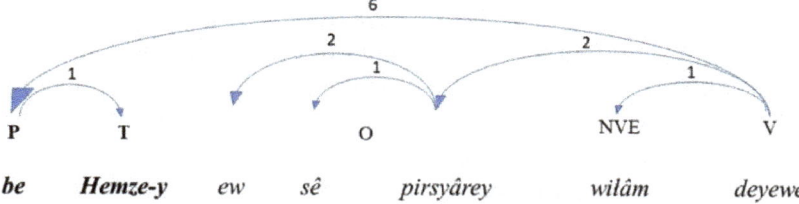

Figure 2: Preverbal Target: total dependency length = 13, IC = 52%.

Although all respondents accept Figure 4, the SPLT and EIC/MiD principles rate it suboptimal, in contrast to Figure 1. The Target is incorporated within the MWV construction together with a relative clause, increasing the dependency length.[24]

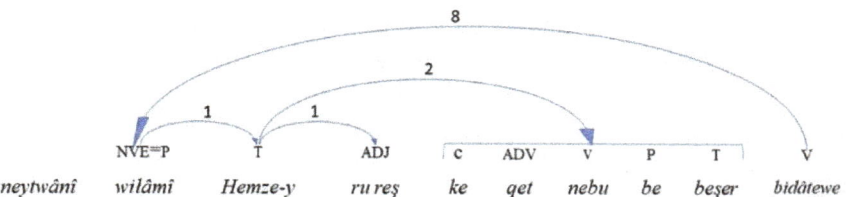

Figure 3: Incorporated Target: total dependency length = 12, IC = 53%.

Finally, respondents consider the example indicated in Figure 5 ungrammatical and it is also not considered optimal based on SPLT and EIC/MiD principles. The reason is that *wiłâm dânewe* is a fixed MWV, which does not allow for any intervening element (Section 2.1).

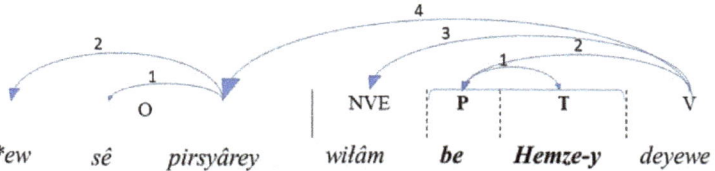

Figure 4: Incorporated Target: total dependency length = 13, IC = 57% (ungrammatical).

---

24 Translation of Figure 4: 'He couldn't answer the shameless Hemze who never came to his senses.'

Thus, comparing the dependency distance and length of the sentences in Figures 1–4, Gibson's principle should prefer the word order of Figure 1 and Hawkins' IC-to-ratio finds Figure 5 is the most optimal structure with Figure 1 the second-best structure. Considering Figure 2, Figure 3, and Figure 4, the dependency length has a minor difference of 1.[25] If the relative clause in Figure 4 was dislocated to the postverbal position, the structure would no longer be problematic for the SPLT principle (dependency length = 8), and it would be the third-best structure for the IC-to-ratio.

Respondents ranked these examples differently than the principles. The crowdsourcing experiment yielded the following hierarchy: Figure 2 > Figure 1 > Figure 4 > Figure 3, and they ranked Figure 1 and Figure 2 nearly identically.

If we assume that an order is optimal if the Target is close to the verb, it would explain the informants' ranking rather well. This is the case in Figure 1 (directly following), Figure 2 (directly preceding) and Figure 4 (incorporated) in contrast to Figure 3. The ranking hierarchy shows that speakers prefer word orders with the closest proximity between the Target and the verb.

While the SPLT principle does not correctly predict the preference of speakers nor the ungrammaticality of Figure 5, it shows that preverbal and incorporated Targets result in similar dependency length, viz. preverbal: *be Hemzey wiłâm dedemewe* [length = 5] 'I give an answer to Hemze' vs. incorporated: *wiłâm be Hemzey dedemewe* [length = 6]. Thus, based on the length of the constructions, incorporation can be considered as a variant of the preverbal construction. The other form of incorporation, using *ezâfe*, seems to be more optimized because flagging by an *ezâfe* creates a syntactic unit between Target and MWV in contrast to flagging by a preposition (e.g. *wiłâmî Hemzey dedemewe* [length = 3] 'I give Hemze an answer'). This is added to the similarity of preverbal and incorporated position in Table 6 in Section 4.

5 Conclusion

Targets can appear between a non-verbal element and the finite verb, which form a MWV construction. In MWVs, three types of non-verbal elements exist in Mukri, viz. 'subject-like' (*bon hâtin* 'wafting smell'), 'object-like' (*wiłâm dânewe* 'to give an answer'), and 'adpositional' (*lêdân* 'to hit'). The subject and object-like types – and the adpositional to some degree – show flexible placement in

---

[25] Note that Figures 1, 3, 4, and 5 were adjusted to test long Target in incorporated positions.

contrast to petrified elements. Especially object-like and adpositional MWVs can adopt new readings so that the compositional semantics shift to a figurative, i.e., petrified, reading. Not all petrified MWV allow Targets to be incorporated even if the structure is considered optimal in terms of processing principles, for instance, if a petrified MWV with a nominal element combines with a direct object (*wilâm dânewe*).

The peculiar feature of the incorporated position is *ezâfe* flagging. Since Targets in the incorporated position can be flagged by either an *ezâfe* or an adposition without any semantic change and differ only in focus, *ezâfe* flagging can be considered a variant of adpositional flagging. In constructions with an obligatory adposition plus a verb (*lêdân*), the adposition cannot be cliticized, while in constructions with a facultative adposition (*lêkutân*) the adposition can appear in its clitic variant. These constructions are classified as simple verbs.

In Mukri, Targets appear in the incorporated, pre- and postverbal positions. Table 7 summarizes the discussed features including dependency length.

Table 7: Extended version of the similarity of preverbal and incorporated positions.

	Position		
	preverbal	incorporated	postverbal
Full flagging	+	+	+
Ezâfe flagging	–	+	–
Reduced flagging	–	–	+
Bare	–	–	+
Literal interpretation	–/+	–/+	+
Dependency length	high	high	low
Animacy	high	high	low

All positions allow for full flagging, while only the postverbal position displays reduced and bare flagging. The postverbal position always yields a literal interpretation, while the interpretation in the incorporated and preverbal positions depends on flagging. Animate Targets are mostly distributed in the preverbal and incorporated positions. Dependency length is high if the Target appears in the preverbal or incorporated positions. The data show that Targets in the incorporated position have more similarities with Targets in the preverbal position, suggesting that the incorporated position can be considered a variant of the preverbal position.

The results of this survey show that the EIC/MiD and SPLT principles cannot account for all of the variation found in the ÖM and TONI corpora. It seems that speakers tend to prefer Targets close to the verb and that the incorporated position is a variant of the preverbal position. However, this result may have to be modified when considering other factors, such as verb types, PoS (e.g., bound morphemes), and discourse-pragmatic factors (e.g., information structure).

Acknowledgments: I am grateful to Thomas Jügel for his very useful comments and discussions about the topics in this study, as well as detailed comments on an earlier draft, which improved both the content and the structure of this paper, as did the one anonymous reviewer of this article. The other two reviewers of the article, Chingduang Yurayong and Kateryna Iefremenko, also provided me with feedback. I thank Donald Stilo and Mike Frechette for useful comments on an earlier draft of the study, as well as Geoffrey Haig, who also kindly shared with me the pre-publication copy of his paper. I take responsibility for any remaining shortcomings.

Abbreviations

1, 2, 3	1st, 2nd, 3rd person
+	present feature
–	absent feature
ADD	additive
ADP	adposition
ADV	adverb
COP	copula
DEF	definite
DEM	demonstrative
DO	direct object
EIC	early immediate constituent
EZ	ezâfe
FUT	future
IMP	imperative
IPFV	imperfective
INDF	indefinite
MiD	Minimizing Domains principle
N	number of token
NEG	negation marker
NP	noun
NVE	non-verbal element

O	object
OBL	oblique
OD	own data
OV	preverbal object
P	preposition
PC	pronominal clitic
PL	plural number
PoS	parts-of-speech
POSS	possessive
POSTP	postposition
POSTV	postverb
PPRF	pluperfect
PRO	pronoun
PROG	progressive
PRS	present
PST	past tense
S	subject
SBJV	subjunctive
SG	singular number
SPLT	Syntactic Prediction Locality Theory
T	Target
V	verb

References

Asadpour, Hiwa. 2021. Cross-dialectal diversity in Mukrī Kurdish I: Phonological and phonetic variation. *Journal of Linguistic Geography* 8(2). 1–12. https://doi.org/10.1017/jlg.2021.1.

Behaghel, Otto. 1932. *Deutsche Syntax. Eine geschichtliche Darstellung.* Vol 4. Heidelberg: Winter.

Biberauer, Theresa. 2017. Probing the nature of the final-over-final condition: The perspective from adpositions. In Laura R. Bailey & Michelle Sheehan (eds.), *Order and structure in syntax I: Word order and syntactic structure*, 177–216. Berlin: Language Science Press. https://doi.org/10.5281/zenodo.1117694.

Bögel, Tina & Miriam Butt. 2012. Possessive clitics and ezafe in Urdu. In Kersti Börjars, David Denison & Alan Scott (eds.), *Morphosyntactic categories and the expression of possession*, 291–322. Amsterdam & Philadelphia: John Benjamins. https://doi.org/10.1075/la.199.11bog.

Faghiri, Pegah & Pollet Samvelian. 2021. The issue of 'separability' in Persian complex predicates. In Berthold Crysmann & Manfred Sailer (eds.), *One-to-many-relations in morphology, syntax, and semantics*, 117–149. Berlin: Language Science Press. DOI: 10.5281/zenodo.4638824

Fisher, Ronald A. 1922. On the interpretation of χ^2 from contingency tables, and the calculation of P. *Journal of the Royal Statistical Society* 85(1). 87–94.

Frommer, Paul Robert. 1981. *Post-verbal phenomena in colloquial Persian syntax*. Los Angeles: University of Southern California Dissertation. (http://digitallibrary.usc.edu/cdm/ref/collection/p15799coll36/id/206554, last accessed September 28, 2020).

Gibson, Edward. 1998. Linguistic complexity: Locality of syntactic dependencies. *Cognition* 68(1). 1–76. https://doi.org/10.1016/S0010-0277(98)00034-1.

Greenberg, Joseph. 1963. Some universals of grammar with particular reference to the order of meaningful elements In Joseph Greenberg (ed.), *Universals of language*, 73–113. Cambridge, MA: MIT Press.

Gündoğdu, Songül. 2018. *Argument-adjunct distinction in Kurmanji Kurdish*. Istanbul: Boğaziçi University PhD Thesis.

Haig, Geoffrey. 2002. Noun-plus-verb complex predicates in Kurmanji Kurdish: Argument sharing, argument incorporation, or what? *STUF/Language Typology and Universals* 55(1). 15–48.

Haig, Geoffrey. 2011. Linker, relativizer, nominalizer, tense-particle: On the Ezafe in West Iranian. In Foong Ha Yap, Karen Grunow-Hårsta & Janick Wrona (eds.), *Nominalization in Asian languages: Diachronic and typological perspectives. Vol. 1: Sino-Tibetan and Iranian languages*, 363–392. Amsterdam & Philadelphia: John Benjamins.

Haig, Geoffrey. 2015. Verb-Goal (VG) word order in Kurdish and Neo-Aramaic: Typological and areal considerations. In Geoffrey Khan & Lidia Napiorkowska (eds.), *Neo-Aramaic and its linguistic context*, 407–425. Piscataway: Gorgias Press.

Haig, Geoffrey. 2017. Western Asia: East Anatolia as a transition zone. In Raymond Hickey (ed.), *The Cambridge handbook of areal linguistics*, 396–423. Cambridge: Cambridge University Press.

Haig, Geoffrey. to appear. Post-predicate constituents in Kurdish. In Yaron Matras, Geoffrey Haig & Ergin Öpengin (eds.), *Structural and typological variation in the dialects of Kurdish*. London: Palgrave Macmillan.

Haiman, John. 1983. Iconic and economic motivation. *Language* 59(4). 781–819.

Hawkins, John. 1994. *A performance theory of order and constituency*. Cambridge: Cambridge University Press.

Hawkins, John, A. 2004. *Efficiency and Complexity in Grammars*. Oxford: Oxford University Press.

Hawkins, John. 2008. An asymmetry between VO and OV languages. The ordering of obliques. In Greville Corbett & Michael Noonan (eds.), *Case and grammatical relations: Studies in honor of Bernard Comrie*, 167–190. Amsterdam & Philadelphia: John Benjamins. https://doi.org/10.1075/tsl.81.08ana.

Jahani, Carina. 2018. Post-verbal arguments in Balochi. Conference presentation at *Anatolia-Caucasus-Iran: Ethnic and Linguistic Contacts* (ACIC), Yerevan University.

Jügel, Thomas. 2017. Syntax of Iranian. In Jared S. Klein, Brian Joseph & Matthias Fritz (eds.), *Handbook of comparative and historical Indo-European linguistics* 41, 549–566. Berlin & Boston: De Gruyter Mouton. https://doi.org/10.1515/9783110261288-035.

Jügel, Thomas. this volume. Word order variation in Middle Iranic: Persian, Parthian, Bactrian, and Sogdian.

Jügel, Thomas & Pollet Samvelian. 2020. Topic agreement, experiencer constructions, and the weight of clitics. In Richard K. Larson, Sedigheh Moradi & Vida Samiian (eds.), *Advances in Iranian linguistics*, 137–153. Amsterdam & Philadelphia: John Benjamins. https://doi.org/10.1075/cilt.351.08jug.

Kemp, Charles, Yang Xu & Terry Regier. 2018. Semantic typology and efficient communication. *The Annual Review of Linguistics* 4(1). 109–128. https://doi.org/10.1146/annurev-linguistics-011817-045406

Korn, Agnes. 2009. Western Iranian pronominal clitics. S. *Orientalia Suecana* 58. 159–171.

Korn, Agnes. this volume. Targets and other postverbal arguments in Southern Balochi: A multidimensional cline.

Megerdoomian, Karine. 2012. The status of the nominal in Persian complex predicates. *Natural Language and Linguistic Theory* 30(1). 179–216. https://doi.org/10.1007/s11049-011-9146-0

Mithun, Marianne. 1984. The evolution of noun incorporation. *Language* 60(4). 847–894.

Öpengin, Ergin. 2016. *The Mukri variety of Central Kurdish: Grammar, texts, and lexicon*. Wiesbaden: Reichert.

Samvelian, Pollet. 2012. *Grammaire des prédicats complexes: les constructions nom-verbe*. Paris: Lavoisier.

Sidwell, Paul. 2020. Nicobarese comparative grammar. In Mathias Jenny, Paul Sidwell & Mark Alves (eds.), *Austroasiatic syntax in areal and diachronic perspective*, 82–104. Leiden: Brill. https://doi.org/10.1163/9789004425606_005.

Stilo, Donald. 2018. Preverbal and postverbal peripheral arguments in the Araxes-Iran linguistic area. Conference presentation at *Anatolia-Caucasus-Iran: Ethnic and Linguistic Contacts* (ACIC), Yerevan University.

Wasow, Thomas. this volume. Factors influencing word ordering.

Agnes Korn
Targets and other postverbal arguments in Southern Balochi: A multidimensional cline

To my sister in Balochi studies, Carina Jahani

Abstract: This article presents a discussion of word order in Southern Balochi. I argue that the postverbal position is the default one for Goals of motion verbs, and also frequently found for Goals of caused motion. Concerning Recipients and Beneficiaries, there is a preference for placing them in "mirror position" with regard to the direct object. Adverbials are also often found in postverbal position, and the same is true for direct objects under certain conditions. These data from an Iranian language spoken in the far south-east of the Iranian sphere provide a point of contrast for studies of word order in the Iranian-Semitic contact zone.

Keywords: Goals and Targets; Iranian linguistics; SOV languages; word order

1 Introduction and framework

1.1 SOV and postverbal arguments

Iranian languages are generally said to be of the SOV type. While this is surely correct as far as the unmarked position of arguments is concerned, it would be a mistake to conclude that the majority of sentences in a given Iranian language shows this order, with noun phrases preceding the verb both for the subject and direct object.

First, sentences with both subject and direct object expressed by a noun phrase are rather rare (Haig 2015: 408); often, entities are introduced into the text one by one, and then referred to by pronouns or by zero, pro-drop being very frequent (see also Section 2.4 below). Secondly, it would likewise be a mistake to think that a basic word order of SOV implies that the verb is sentence-final (thus Haig 2015: 413 for Kurdish). There is a considerable amount of variation in actual sentences, and various arguments can be found in postverbal position for pragmatic or other reasons.

Agnes Korn: CNRS; Centre de recherche sur le monde iranien (UMR 8041), 27, rue Paul Bert, 94204 Ivry-sur-Seine, France, E-mail: agnes.korn@cnrs.fr

https://doi.org/10.1515/9783110790368-005

1.2 The concept of Target

In Iranian languages, the postverbal position is particularly frequent for nominal or adpositional phrases, adverbs etc. expressing the end point of a movement or action. This preference can metaphorically be seen as an iconic principle in the sense that the action/event expressed by the verb is directed to a certain end point.[1]

This paper will use the term "Target" as an umbrella term for various types of arguments having in common that the action or event aims towards a certain end point. The types of non-subject[2] Targets which will be studied are those in (1). The inclusion of "Final state" follows Jahani (2018) and Asadpour (this volume).

(1)   Targets studied in this paper
    a.   Goals of verbs of motion (*go **to the market / to grandmother's place***),
    b.   Goals of verbs of caused motion (*bring the cattle **home / to the landowner***),
    c.   Recipients of verbs of transfer (*give a book **to the child***),
    d.   Beneficiaries (*do something **for Tim***),
    e.   Addressees of speech verbs (*say something **to the children***),
    f.   Final state (*turn* (transitive or intransitive) ***into a stone***).

1.3 Postverbal arguments and language contact

The starting point for this article is provided by Geoffrey Haig's work on the topic of postverbal elements in SOV-languages, specifically in Kurdish. He states (Haig 2015: 408) that Kurdish combines the features

(2)   OV word order, prepositions and noun-genitive order.

According to a typological study by Dryer (2013), this combination is very rare cross-linguistically, but "frequent" in the contact zone of Iranian and Semitic, occurring as it does in, e.g., Persian, Tajik, Kurdish – all members of the Western sub-branch of Iranian – and some North-Eastern Neo-Aramaic varieties (Haig 2015: 408).

---

[1] See Haig (2015: 414); Haiman (1983), etc.
[2] This excludes arguments in subject position that semantically could be seen as a Target in the sense of the definition in (1), e.g. ***The child*** *receives a book for its birthday*, and experiencer constructions (including patterns such as *To me the head is aching*). For Balochi constructions with "non-canonical" subjects, see Jahani et al. (2012).

Haig (2015: 408f., following Harris and Campbell 1995: 137) states that the configuration in (2) is one of the features that "only arise in contact situations" such as shown by Persian, Tajik and Kurdish, whose position as "outlier" of an (otherwise) "OV / postposition / G[enitive]N[oun] block" (which includes "basically the entire Asian land block eastwards of Anatolia") exposes them to contact with languages from other families.

According to Haig (2015: 414), language contact not only causes the configuration in (2) but may also motivate the occurrence of Targets in postverbal position in languages of the region: Both Kurdish and North-Eastern Neo-Aramaic show a combination of OV and postverbal Goals, which is unusual for both (head-initial) Semitic and historically (head-final) Iranian.[3] It is this configuration of language contact, says Haig (2015: 420–421), which explains the difference between postverbal Goals such as 'go home / to grandmother's place' being regular for some Kurdish varieties and less generalized in Kurdish varieties outside the Iranian-Semitic contact zone.

1.4 Providing a point of comparison for the Iranian-Semitic contact zone

The research perspectives outlined in Section 1.3 have inspired a number of publications on postverbal arguments in the Iranian-Semitic contact zone, this volume being an example. At the same time, it seems to me that a focus on shared features in a zone of contact is at risk of overstating its point if it is not contrasted with languages outside the area. Indeed, the status of postverbal arguments (if any) in languages not in the Iranian-Semitic zone has not been much studied yet, hence little contrastive material is available. The present article discusses material from Balochi, an Iranian language spoken outside this contact zone, i.e., not in contact with Semitic in recent times (or perhaps not at all). At the same time, Balochi belongs to the Western Iranian branch, several other members of which take part in the mentioned contact zone (see Section 1.3). This paper thus discusses Targets and their position (Section 3) and then compares these to other elements in postverbal position (Section 4).

---

3 Haig's study includes Goals of verbs of motion; recipients of verbs of transfer; addressees of speech verbs (Haig 2015: 413). Jahani (2018) includes Goals of verbs of motion, Recipients of verbs of transfer, Addressees and Final states.

Map 1: Balochi dialects (red frame: region where the examples come from).

2 Data and approach

2.1 The position of Balochi

Balochi is a Western Iranian language with the maximum distance to the Iranian-Semitic contact zone today and indeed situated more in the Indo-Aryan sphere. It largely shows the left-branching structures mentioned in Section 1.3, i.e., genitives preceding nouns and postpositions (largely relational nouns in the oblique case, see Section 2.4). NPs and PPs are thus of a structure exactly opposite to Persian. In reality, matters are somewhat more complicated, though, surely under the influence of the respective national languages. The basic pattern combines a few inherited prepositions (chiefly *ač/ča* etc. 'from', a cognate of Persian *az*, and *pa(r)* 'to/for') with lexical postpositions ('on', 'in', 'behind'), but some dialects in Iran have acquired more right-branching features, while the dialect of Karachi in Pakistan shows an entirely postpositional pattern (Jahani and Korn 2009: 657).

In spite of its present habitat, Balochi is likely to have come from the Caspian Sea area, and shares features with languages spoken there.⁴ In addition, it seems likely to me that, during its move from the Caspian Sea to its present location, speakers came into contact with Kurdish. This may be seen in sound changes that both Balochi and Kurmanji share with Persian (Korn 2019: 247) and by animal terms, some of which may have been borrowed from Kurdish into Balochi and at least one item the other way round.⁵ It seems that by the time of the Arab conquest, speakers of Balochi had moved away and arrived in the Kerman region. The presence of Baloch in Pakistani Balochistan and adjacent regions might date back some 500 years (see Korn 2005b: 43–45). Balochi being spoken in more western and northern regions is probably the result of more recent (back-)migrations. The westernmost dialect is Koroshi, spoken in the province of Fars.

2.2 The data used for this article

The data chosen for this paper are from recordings made by Maryam Nourzaei among the "Afro-Baloch" in the Southern Balochi speaking area of Iranian Balochistan.⁶ Compared to Southern Balochi dialects spoken across the border in Pakistan on the one hand and to other Balochi dialects in Iran on the other, the speech of the "Afro-Baloch" is comparatively "archaic", e.g., as far as the nominal system or ergative patterns are concerned. This is fortunate for the present purposes since there is a tendency for minority languages in Iran to replicate Persian patterns. In what follows, "Balochi" refers to these data unless otherwise specified.

Note that with the focus on the data just mentioned, the present study is not representative of Balochi as a whole. It is also based on a survey of the data rather than on a quantitative analysis. The results will thus be somewhat preliminary and might raise at least as many questions as this paper attempts to answer. The idea is, however, to provide comparative material to be contrasted with the languages spoken in the Iranian-Semitic contact zone.

---

4 See Boyajian (2003) for data from historical and literary sources, Korn (2005a: 299) for features in the nominal system shared with Gilaki, and Korn (2016: 409–411) for a phonological isogloss of both languages. At the same time, Balochi has of course also been under the influence of Persian. I argue that, rather than continuous influence, there were waves of influence coinciding with the periods of Persian as national language (Korn 2019: 246); this is thus not a case of geographical proximity, but of the effect of a superstrate.
5 Cf. Korn (2006).
6 A grammatical survey of these data and a description of their analysis can be found in Korn and Nourzaei (2019).

2.3 Presentation of the data

All the examples in this article are from the recordings mentioned in Section 2.2.

As will become apparent, the textual structure is quite important for word order. A number of examples will thus be cited with preceding or following sentences. The second purpose of citing examples in context is to illustrate which elements other than Targets may occur in postverbal position. For this reason, finite verbs (including light verbs of complex predicates) are marked in bold in all of the sentences cited below for a visual impression of word order in Balochi in general.

Single quotation marks are used at the beginning and the end of the quoted passages. These may consist of several sentences, which are numbered individually to permit referring to them in one or the other subsection. Double quotation marks are used for quoted speech within the examples.

Curly brackets {} will be used in the translation of examples to highlight the element discussed in the section concerned.

Examples published in Korn and Nourzaei (2019) will be cited as "K&N" with example number. Other examples will give the abbreviation of the place of the recording and the initials of the speaker, followed by the number of the text and the sentence.[7] The text "Ko[narak] GO 1" is published in Korn and Nourzaei (forthcoming). Glosses have been slightly unified.

I will make occasional reference to Barjasteh Delforooz (2010), which contains some discussion of word order in Balochi of the Iranian part of Sistan, belonging to the so-called Western Balochi dialect group and showing more pronounced Persian influence. I will also make use of a paper on word order presented by Carina Jahani in Yerevan in 2018, which she kindly gave to me for further use.

2.4 Some notes on definitions and grammar

Concerning the marking of arguments in Balochi, all types of Targets listed in (1) normally use the **oblique case** (SG -$ā$/-a, PL -$ān$). Its functions include marking direct and indirect objects as well as the agent in transitive constructions in the ergative domain, but direct object marking is limited to definite direct objects in the nominative domain. Pronouns have a separate object case, which is preferred to the oblique when the pronoun is the direct or indirect object.

---

7 See the abbreviations at the end of the article.

The oblique case is also used for locations and directions in time and space (including Goals of verbs of motion and caused motion) as well as for various adverbial functions ('in the house', 'at night', 'in this way', etc.).

For more precision, Balochi uses **postpositions** (cf. Section 2.1), which largely are relational nouns in the oblique case, e.g., *tah-ā* 'in' (lit. 'in the interior') and the preceding noun takes the genitive case. Just like any noun in the oblique case, the adpositions can also be used alone as adverbs. Prepositions usually take the oblique case of the noun.

As will be discussed in Section 6.4, **local reference to humans** is of different types in the data. All of them involve the genitive and the oblique:

X-GEN-OBL, lit. 'at (OBL) X's (GEN) [place]' (rare);[8]
X-GEN *gwar-ā* lit. 'at (OBL) X's (GEN) side (*gwar*)' and other adpositions;
X-GEN *lōg-ā* 'at (OBL) X's (GEN) house (*lōg*)' and similar expressions (frequent).

Given the fact that the only option to locally refer to a person is by possessive constructions, all three patterns will be considered as human reference for the purposes of this article, although the third pattern of course also refers to X's house (or castle, etc.) at the same time.

Somewhat disadvantageous for a study on word order, Balochi is a heavily **pro-drop**ping language (one might even say "argument-drop"): Elements that native speakers can retrieve from the context (for a linguist, this is not always easy) tend not to be mentioned in the sentence, reducing potential examples of certain types.

Another reduction of possible examples occurs by the very frequent use of **pronominal clitics** (enclitic pronouns, PC), which can be used for all functions of the oblique case, including some Targets (Recipients, Beneficiaries and Addressees). Clitics may be assumed to follow placement rules different from those of NPs, and will not be studied in this article.

Balochi very frequently uses **combinations of several verbs**, which can be of various types. Applying the definitions by Bashir (2008: 65–75), serial verbs are combinations of several inflected verbs joined without a connector such as 'and', they can refer to one or several events. Combinations of a verb with a semantically bleached "vector verb" refer to a single action, and the vector verb "contributes aktionsart or aspectual meanings" (Bashir 2008: 74). The latter is

---

8 Cf. Korn and Nourzaei (2019: 628–630). Some Balochi dialects of the Western group have developed this combination of endings into a locative case, see Korn (2008). This is not to be confounded with the combination of the individuation marker and the oblique case, which is found in (8), (24) and (75).

the case of the series 'seize, come' meaning 'bring back' in (90) or of 'bring, seize' in (43), where the 'seize' would be in the wrong position if the sentence were to be taken literally. Owing to pro-drop etc., the different types of combinations are not always easy to distinguish, and "constitute a continuum of grammaticization of verbal conceptions" (Bashir 2008: 66).[9]

The passages cited as examples will highlight that **Tail-Head-Linkage** (THL, see Section 6.2) is quite common in Balochi. THL refers to the repetition of (some of) the preceding clause (verbatim, with some changes or in an abbreviated form) before proceeding to new material. In this way, the repeated part serves as a background to what follows and assures text coherence.[10] The repeated part can often be interpreted as equivalent to a subordinate clause, see Korn (2017).

3 Postverbal Targets

As pointed out in Section 1.3, postverbal Targets are common in languages of the Iranian-Semitic contact zone, and certain subtypes have even been generalized in some varieties. For an article seeking to study postverbal elements in an Iranian language outside this zone, the position of Targets is thus the first issue to be tested. This section will discuss the pre- and postverbal placement of Targets in the Balochi data described in Section 2.2 and possible conditions for such placement.

Combinations of verbs of motion and transfer are frequent in the data. As it will turn out, some of the verbs which may be expected to be used with a Target hardly ever do so; see Section 6.1 for discussion of these points.

3.1 Verbs of motion: 'go', 'come', 'arrive'

As pointed out by Haig (2015: 413), "there are reasons to distinguish" human and non-human Targets of verbs of motion; this section is thus arranged in this way.

---

9 See also Korn and Nourzaei (2019: 647–649). See Section 6.1 for more discussion.
10 See Barjasteh Delforooz (2010: 47, 102, 111–120, 280) and Nourzaei (2017: 68, 165–169, 201f., 255–257, 305f.) for THL in Balochi.

3.1.1 [–human]

Postverbal position

Goals of motion verbs are very frequent, and they are commonly found in postverbal position. Further example: (23).

'go'

(3) č-edā   ē   **laḍḍe**   šot-ã   gōdēngar-a
    from-here   PROX   move.PST   go.PST-3PL   PN-OBL
    'From here, they moved [and] went {to Godengar (a village near Konarak)}.'
                                                              [Ko SE 1/3:3]

(4) **gwašt**   **b-ra-∅**   ham-ō   kahīr-a
    say.PST   IRR-go.PRS-IMP.2SG   EMPH-DIST   PN-OBL
    '[My mother] said: "Go {to that Kahir (village)}..."'   [Ko SE 1/2:27]

(5) o   wat-ā   **larz-ēnt=ī**   **šo**   lōg-ā
    and   self-OBL   shake-CAUS.PST=PC.3SG   go.PST.3SG   house-OBL
    '... and he (the lion) shook himself [and] went {home}.' [Ko GO 1.94f.]

(6) **zort=ī**   maanā-ēn   sarīg   **šo**   ham-odā
    seize.PST=PC.3SG   silk-ATTR   headscarf   go.PST.3SG   EMPH-there
    'She took the silken headscarf [and] went {there}.'   [Shi YN 1.195]

(7) asabī   **kot=et**   mardom-ān   **be-rr-et**   dar-ā
    nervous   do.PST=PC.2PL   people-OBL.PL   IRR-go.PRS-2PL   out-OBL
    (to the children:) 'You are upsetting people (you are disturbing the recording); get {out} (of here)!'   [Ka GH 1.49]

(8) mē   mā   **šo**   sīr=ē-a
    we.GEN   mother   go.PST   wedding=IDV-OBL
    'Our mother went {to a wedding}.'   [Ka AJ 1.6]

(9) o   mē   mā   ēdga rōč-a   **šed-a**   sīr-a
    and   we.GEN   mother   other day-OBL go.PST-PRF.3SG   wedding-OBL
    'And our mother went {to the wedding} [again] the next day.'[11] [Ka AJ 1.13]

(10) **šot-a**   šakār-ā
     go.PST-PRF.3SG   hunt-OBL
     'He (the boy) went {hunting}.

---

11 Balochi wedding celebrations last several days.

(11) rāh=ī   dā   b-ro-∅   šakār-a
     way=PC.3SG give.PST   IRR-go.PRS-IMP.2SG hunt-OBL
     He (the king) sent [him and said]: "Go {hunting}!"

(12) rāh=ī   dā   ēšī   šakār-ā
     way=PC.3SG give.PST   PROX.OBL   hunt-OBL
     He (the king) sent him {hunting}.

(13) šo   šo   šakār-a   ǰangal-ā   šo
     go.PST.3SG   go.PST.3SG   hunt-OBL   forest-OBL   go.PST.3SG
     He (the boy) went [off and] went {hunting}; he went {to the forest}."[12]
     [Shi YN 1.10-14]

'come':

(14) ā   ya   žapaga   tambū-ē   dapā
     DIST   come.PST.3SG   quickly   tent-GEN   in_front_of
     'He quickly came {to in front of the tent}.'   [Shi YN 1.38]

Preverbal position

The preverbal position is less common for the Goal of a verb of motion, but it does occur. Further example: (13).

(15) ham-ē   bayt-a   gwaš-ān   kot   o
     EMPH-PROX   verse-OBL   say.PRS-PTCP   do.PST   and
     dēm   pa   zīyārat-a   raw-ān
     front   for   shrine-OBL   go.PRS-PTCP
     'They said this verse and went {to the shrine}.

(16) ōdā   šot-an
     there   go.PST-3PL
     They went {there}.'   [Shi YN 1.255f.]

(17) gwat=ī   ham-ā   daryā   kap-e
     say.PST=PC.3SG   EMPH-DIST   sea   fall.PRS-2PL
     ke   ǰōl-ter   en
     SUB   deep-EL   COP.PRS.3SG
     'He (the lion) said: "Fall {into that sea}, which is very deep!"'
     [Ko GO 1.126f.]

(18) šot   šot   šot   šot   šot
     go.PST.3SG   go.PST.3SG   go.PST.3SG   go.PST.3SG   go.PST.3SG
     šot   šot   šot   šot
     go.PST.3SG   go.PST.3SG   go.PST.3SG   go.PST.3SG

---

[12] The hunt(ing) in this passage is a purpose-like Target, see Section 4.1.

	šap	ǰat-a			
	night	hit.PST-PRF			

'He went, went, went, went, went, went, went, went [until] night fell.'

(19) ham-ā molk-ā ke šo ...
EMPH-DIST country-OBL SUB go.PST.3SG
'When he came {to that country}, ...' [Shi YN 1.128f.]

3.1.2 [+human]

[+human] Goals of motion verbs are much less frequent. They are likewise found in postverbal and preverbal position, the former being more frequent. (27) shows "entering" a body part. Further example: (52).

'go':

(20) mard-ā **hakalet**
man-OBL leave.PST
šo pada ot-ī padešāh-e gwarā
go.PST.3SG back REFL-GEN king-GEN by
'The man left and went {back to his king}.'

(21) bādšāh-e gwarā ke **šo**
king-GEN by SUB go.PST.3SG
'When he came {to the king}, [the king asked: ...]' [Shi YN 1.119f.]

(22) **šo** ham-ā ballok-ē lōg-ā
go.PST.3SG EMPH-DIST grandmother-GEN house-OBL
'He went {to the old woman's house}.' [Shi YN 1.200]

'arrive':

(23) ǰenek **šo** gapaṭ-ok=ē dā ōtī molk-ā
girl go.PST.3SG distance-DIM=IDV until own country-OBL
'The girl went a little distance {to her country}.'

(24) gapaṭ=ī-a **šo** **šo**
distance=IDV-OBL go.PST go.PST.3SG
dīt=ē ya dokter=ē hesāb ka-∅
see.PST=PC.3SG one doctor=IDV reckoning do.PRS-IMP.2SG
'[As] she went a little distance, she saw a doctor, imagine.[13]'

---

[13] hesāb ka 'imagine!' is addressed to the audience.

(25) tarā   ma   hazār   kaldār   a   **dey-õ**
     you.SG.OBJ I   thousand   (money)   VC   give.PRS-1SG
     [She said to him]: "I'll give you a thousand ducats.

(26) sāl-a   tarā   hazār   **ras-ī**
     year-OBL you.SG.OBJ   thousand   arrive.PRS-3SG
     Next year[14] [another] thousand will come {to you} (if the boy recovers, I will give you more).'" [Shi YN 1.108-112]

(27) dā̃ke   sād   **potret**   m-ē   badan-ē   tōkā
     until   rope   enter.PST.3SG   EMPH-PROX   body-GEN   in
     '... until the rope cut (lit. entered) {into his body}.'   [Ko GO 1.47]

3.1.3 Discussion

Goals of motion verbs are highly frequent, but the only instance of the very frequent 'come' in the data is (14), with the Goal in postverbal position (see Section 6.1 for further discussion).

The postverbal position seems to be the default one for Goals of motion verbs, particularly when the Goal is [–human]. They are commonly found without adposition (see Section 2.4). (3) is a clear instance of the iconic principle mentioned in Section 1.2: The starting point of the movement comes first, then the two verbs, and the end point of the action terminates the sentence.

However, preverbal Goals are found as well. One might assume that length plays a role, i.e., that an element consisting of only one word would tend to precede the verb and a long NP or PP rather follow it. In this sense, [+human] Goals, given that they are often used with an adposition or expressed as 'at X's house' (see Section 2.4), might tend to be postverbal. However, there are sentences with parallel post- and preverbal Goals, cf. (6) vs. (16), or (3) and (4) vs. (15). A quantitative study would be necessary to show whether length plays a role for Goals of motion verbs.

Conversely, there is a clear tendency for a preverbal Goal referring to a place already mentioned. Quite typical are passages with a Goal placed after the verb first, and re-occurring in the subsequent text (often in the next sentence) in preverbal position, as is the case in (21) and (24). The 'forest' in (13) and the 'country' in (19) could also be interpreted as being implied in the preceding text.

---

14 It is not clear from the context whether this means: next year, or: every year to follow.

The reverse does not apply: In (9) and (11)–(13), the wedding and the hunt remain postverbal in spite of having been mentioned in the preceding sentence. Note that the wedding has the "individuation marker" =ē in (8) and none in (9), and both are treated alike, confirming that the postverbal position is the default one independently of such marking.

3.2 Verbs of caused motion: 'bring', 'carry', 'transport', 'send' and 'put'

Verbs of caused motion include 'bring', 'carry', 'send' and the like. Also included here are expressions broadly meaning 'put'; these use 'do', 'give' etc., often in combination with local adverbs or adpositions.

[–human] Goals of caused motion verbs are frequent only for expressions meaning 'put'. [+human] Goals of caused motion are moderately common. The preverbal position seems at least equally common as the postverbal one, with Goals of equal length.

3.2.1 Postverbal position

[–human]
Further example: (12).

(28) **bort**=e   ōtī   molk-ā
      carry.PST=PC.3SG   own   country-OBL
      'He brought [her] {to his own country}.'

(29) ōdī   **bo**   bādešāh   **go**
      that_place   carry.PST   king   say.PST
      [As he] brought [her] {there}, the king said: ...'   [Shi YN 1.260f.]

(30) nī   **bar-∅=ī**   drang-ā
      now   carry.PRS-IMP.2SG=PC.3SG   rack-OBL
      (speaking about the goat skin containing the milk): 'Now, bring it {to the rack}.'   [Ko SE 2.41]

(31) hord-olok-ḗ   poštok   a   **kan-ḗ**
      small-DIM-ATTR   ball_of_dough   VC   do.PRS-1PL
      'We make small dough balls

(32) o   **bar-ēn**=ē   ham-ē   ǰām-ok-ā̃
      and   carry.PRS-1PL=PC.3PL   EMPH-PROX   pot-DIM-OBL.PL
      and we place them {into a small pot}.'   [Ka GH 1.73f.]

(33)　o　**k-ār-en**
　　　and　k.IND-bring.PRS-1PL
　　　o　**rēč-ēn**　　ham-ē　　　lōhī-yē　lāfa
　　　and　pour.PRS-1PL　EMPH-PROX　jar-GEN　inside
　　　'... and we bring [it] and pour [it] {into this jar}.' [Ka GH 2.110]

[+human]
(35) and (47) contain body parts.

(34)　man　tarā　　**ārt-a**　　　　pa　ōtī　bādešāh-ē-a
　　　I　　you.SG.OBJ　bring.PST-PRF　for　own　king-GEN-OBL
　　　'I brought you {to my king}.' [Shi YN 1.267; K&N 2]

(35)　ōtī　ǰanēn=ī　　　edān　**koṯ-a**　　　mā̃　koṯ-a
　　　own　woman=PC.3SG　here　do.PST-PRF　in　lap-OBL
　　　'He put his wife {here on his lap}.' [Ko SE 1/1.57]

3.2.2 Preverbal position

[–human]
Further examples: (29) and 'here' in (35).

(36)　ēšī-yā　　　　ma　har-ǰā　　　**bar-ē̃**
　　　PROX.OBL-OBL　we　every-place　carry.PRS-1PL
　　　ē　　　**ǰerg-ī**
　　　PROX　escape.PRS-3SG
　　　'... {everywhere} we bring him, he escapes.' [Ko SE 1/1.106f.]

(37)　mā　pād-a　　**k-ō-ēn**　　　　　o
　　　we　foot-OBL　k.IND-come.PRS-1PL　and
　　　mā̃　ōtī　nwād-a　　　sara　**šān-ēn**
　　　in　own　bedding-OBL　up　　throw.PRS-1PL
　　　'We get up and vomit {into our bedding}.' [Ka AJ 1.26]

(38)　sēom　rōč-a　　ēr　　a　　**gēǰ-ē̃**　　　　**k-ār-ēn=ē**
　　　third　day-OBL　down　VC　throw.PRS-1PL　k.IND-bring.PRS-1PL=PC.3PL
　　　o　　langarī=yē　**kan-ēn=ē**
　　　and　tray=IDV　　do.PRS-1PL=PC.3PL
　　　'On the third day, we take [it (the yoghurt)] down and bring it (lit. them),
　　　and we place it {onto a tray}.'[15] [Ka GH 1.59f.]

---

[15] Milk products, tea, water etc. are considered as plural in Balochi, thence the 3PL pronominal clitic here and in other examples.

(39) ēr   **rēč-ḗ**   šīr-ān
     down   pour.PRS-1PL   milk-OBL.PL
     mā̃   lōhī-hok-ān   a   **kan-ēn**=ē
     in   pot-DIM-OBL.PL   VC   do.PRS-1PL=PC.3PL
     'We pour the milk [and] put it {into the pots}.'   [Ka GH 1.38]

(40) o   mā   balāh-ḗ   tālom=e   sāṛō=īy
     and   we   huge-ATTR   tray=IDV   big_bowl=IDV
     a   **kan-ēn**=ē
     VC   do.PRS-1PL=PC.3PL
     '... and we put it {into a huge plate [or] bowl}.'   [Ka GH 1.123]

(41) o   šīrēnč-a   **by-ār-∅**
     and   funnel-OBL   IRR-bring.PRS-IMP.2SG
     ma   hīzak-e   dap-ā   **dey-∅**
     in   goatskin-GEN   mouth-OBL   give.PRS-IMP.2SG
     'And bring a funnel and put [it] {into the mouth of the goatskin}.'
                                                                [Ko SE 2.37]

(42) o   degar=ē   mā̃   lačok-ē   lāfā   **kan-ḗ**
     and   then=PC.3SG   in   basket-GEN   inside   do.PRS-1PL
     '... and after that we place it {in a basket}.'   [Ka GH 1.65]

(43) **ārt-ag-ant**=ē   o   **zort-ag-ant**=ē
     bring.PST-PRF-3PL=PC.3PL   and   seize.PST-PRF-3PL=PC.3PL
     o   rōč-a   **kod-ag-ant**
     and   day-OBL   do.PST-PRF-3PL
     'They (= the poor people) took it (home), and they put it {in the sun}.'
                                                         [Ka GH 2.22; K&N 51]

(44) o   **bar-ēn**   ās-a   **kan-ēn**
     and   carry.PRS-1PL   fire-OBL   do.PRS-1PL
     ' ...and we go and put [it] {onto the fire}.'   [Ka GH 1.44]

(45) čē   kambol-ān   a   **kan-ag=ant=ī**
     down   cover-OBL.PL   VC   do.PRS-INF=COP.3PL=PC.3PL
     'We put it {under blankets}.'   [Ka GH 1.10; K&N 29]

[+human]
Further example: (61).

(46) āy-ānī   lōg-ā   warag-ā̃   **bar-ant**
     DIST-GEN.PL   house-OBL   food-OBL.PL   carry.PRS-3PL

	talā	*čo*	**rasēn-an**	**k-āy-an**	
	moment	like	transport.PRS-3PL	k.IND-come.PRS-3PL	

'They (= the relatives, or servants) deliver food {to their house (= bride & groom)} quickly and then return.' [Shi YN 1.249]

(47)	*o*	*edā*	*koṭṭ=ey*	**p-kan-et**	
	and	here	lap=PC.3SG	IRR-put.PRS-2PL	
	o	**b-zīr-et**=*ī*		**b-r-e**	
	and	IRR-seize.PRS-2PL=PC.3SG		IRR-go.PRS-2PL	

'... put him {here on your lap} and take him [and] go.' [Ko SE 1/1.19]

3.2.3 Discussion

Some examples of verbs of caused motion behave similarly to verbs of motion, e.g., 'carry' in (28), featuring the postverbal Goal 'to his country' (which was already mentioned earlier in the text) and a preverbal one in the next sentence (29), parallel to 'go' in (20)–(21) above. Example (34) recalls (20), the [+human] reference expressed here by the combination of case markers mentioned in Section 2.4 in addition to the preposition, while (46) has 'their house'.

However, while postverbal Goals of caused motion verbs are certainly not rare, the preverbal position seems to be no less common.

More importantly, instances of [–human] Goals of caused motion verbs other than those meaning 'put' vel sim. are conspicuously infrequent. Specifically for 'bring', it seems that the situation is parallel to 'come', which does not normally take an overt [–human] Goal.

In addition, it seems that while Goals of 'go' typically are [–human], viz. the place where someone goes to (see Section 3.1.1), those of caused motion tend to express the person (the person's place) to whom an object is taken, thus [+human]. This renders such Goals similar or even identical to Recipients (see Section 3.3). Still, one cannot say that verbs of motion with a [+human] Goal abound.

Expressions for 'send' are likewise rare. The passage (10)–(13) does contain one 'send (hunting)' with postverbal Goal (12), and (61) has a preverbal Goal while *rasēn-* 'transport' in (46) is without Goal.

Conversely, expressions meaning 'put' etc. quite commonly have a Goal. They usually contain 'do', less frequently 'give' (41) or 'carry' (32), with or without an adverb or nominal element.[16] There seems to be a free variation as to the

---

16 German shows parallels for such uses, mostly in combination with preverbs: Austrian German regularly uses *geben* 'to give' (e.g., *X hinter Y geben* 'to put X behind Y', *hinein-geben*

position of the Goal, cf. (33) vs. (42), both with a container marked by the postposition *lāfā/lafa* 'inside', or (32) vs. (38), (40) etc. without adposition. The preverbal position seems to be more frequent, even with a particularly long Goal such as the one in (40). These expressions thus seem to pattern differently than verbs of caused motion such as 'bring' and 'carry'.

3.3 Recipients and beneficiaries

Somewhat related to Goals of caused motion verbs are Recipients of verbs of transfer and Beneficiaries, in the sense of representing the end point of the action or someone being transported to a certain person.

3.3.1 Postverbal position

The postverbal position is common, particularly when the Recipient / Beneficiary is mentioned for the first time.

(48) sad   tomon=ī   **dā**   ballok-ā
     hundred   (money_unit)=PC.3SG   give.PST   grandmother-OBL
     'He gave a hundred Toman {to the old woman}'   [Shi YN 1.153]

(49) dega   bahr   a   **kan-ē**   pa   ōtī   hamsāheg-ān
     then   share   VC   do.PRS-1PL   for   own   neighbor-OBL.PL
     'Then we make shares {for our neighbors}

(50) o   gōṭok=ē   gōṭok=ē   ōtī   hamsāheg-ān   a   **day-ē̄**
     and   rest=IDV   rest=IDV   own   neighbor-OBL.PL   VC   give.PRS-1PL
     and we give a little, a little {to our neighbors}.'   [Ka GH 110f.]

(51) čē   ham-edān   kāgad=ē   **dāt-a**   mnī   pet-a
     from   EMPH-here   letter=IDV   give.PST-PRF   I.GEN   father-OBL
     'From here he (Golmahmad) gave a letter {to my father}
     hoseyhān-a   če   ham-ē ...   golmahmad-a   golmahmad-a
     PN-OBL   which   EMPH-PROX   PN-OBL   PN-OBL
     {for Hoseinkhan}, what [was it] ... this Golmahmad, Golmahmad.

(52) **dāt-a**   kāgad   šot-a   hoseyhān-ī   kalāt-a
     give.PST-PRF   letter   go.PST-PRF.3SG   PN-GEN   estate-OBL

'to put into', *hinauf-geben* 'to put onto'), and *machen* 'to do, produce' is found in colloquial German of Germany (e.g. *hinein-machen* for 'put into (a container)') as well as *tun* 'to do', while the standard language uses *setzen, stellen, legen* (all 'to put, place').

[When Golmahmad] gave the letter, he (my father) went to Hoseinkhan's village.' [Ko SE 1/1.1-3]

(53) o   ā   mardom=ē   ...
     and   DIST   man=PC.3SG
     ā   ǰanēn=ē   **dāt-a**   gerāk-a
     DIST   woman=PC.3SG   give.PST-PRF   customer-OBL
     'And he [had given] the man... He had given that woman {to a tradesman}

(54) ǰanēn-a   gerāk   **bort-a**
     woman-OBL   customer   carry.PST-PRF
     [so that] the tradesman had taken the woman away.' [Ko SE 1/4.33f.]

3.3.2 Preverbal position

Quite typically, 'give' is combined with a verb of motion or caused motion; the recipient then precedes 'give'. (55) shows that the preverbal position is also found without a verb of motion. (59)–(60) contains one Beneficiary and one Recipient.
Further examples: (25), (50).

(55) mā   ēšī-ā   **dah-ḗ**   čelīm-ā
     we   PROX.OBL-OBL   give.PRS-1PL   waterpipe-OBL
     'We will give {him} a waterpipe.

(56) **k-ār-ī**   šomā-r=ē   **dā**
     k.IND-bring.PRS-3SG   you.PL-OBJ=PC.3SG   give.PRS.3SG
     He will bring [it and] give it {to you}.' [Ko SE 1/1.15f.]

(57) gwašt=ī   **ba-ra-∅**
     say.PST=PC.3SG   IRR-go.PRS-IMP.2SG
     ham-ā   bānū-r=ē   **be-de-∅**
     EMPH-DIST   bride-OBJ=PC3SG   IRR-give.PRS-IMP.2SG
     'He said: "Go [and] give [it] to {that bride}."' [Shi YN 1.155]

(58) **bar-∅=ī**
     carry.PRS-IMP.2SG=PC.3SG
     ham-ā   gerāk=ī   **be-de-∅**
     EMPH-DIST   tradesman=PC.3SG   IRR-give.PRS-IMP.2SG
     'Take it [and] give it {to that tradesman}!' [Ko SE 1/4.5]

(59) ma   henčo   petāī   par   to   **kot-a**
    I   that_much sacrifice for   you.SG do.PST-PRF
    'I did so much sacrifice {for you}
(60) to   manā   **ārt-a**   ōtī   bādešā-yā   **de**
    you.SG I.OBJ bring.PST-PRF   own   king-OBL   give.PRS.2SG
    [and] you brought me [here and] are [now] giving [me] {to your king}?!?'
    [Shi YN 1.271f.]

3.3.3 Discussion

Both the post- and preverbal position are common for Recipients and Beneficiaries. When both the direct object and the Recipient / Beneficiary are present as an NP, there seems to be a tendency for them to be placed on opposite sides of the verb, with a preference for the Recipient to be placed in postverbal position. The opposite configuration is found, too, cf. (55).

New information tends to be placed after the verb and given information in front of it, as in (53)–(54). This may override the preference for having direct object and Recipient / Beneficiary on opposite sides of the verb, cf. (50), where both Beneficiary and direct object are given information (cf. 49) and are preverbal.

The latter feature recalls the placement of Goals of motion and caused motion verbs. However, while these have a preference for the postverbal position, it seems to me that there is no clear default position for Recipients and Beneficiaries as such, and that it is the factors just mentioned which govern their placement.

3.4 Addressee of speech verbs

The Addressee of a speech verb is usually obvious from the context, so that it is rarely expressed in Balochi, which prefers limiting arguments to the strictly necessary (cf. Section 2.4), cf. the numerous examples in this article of 's/he said' without Addressee.

The addressees that do occur are all in preverbal position in my data.

(61) *hamrāh-ān=ī*   **gwašt=ī**   šām
    companion-OBL.PL=PC.3SG   say.PST=PC.3SG   dinner
    sōbāreg   warag   ballok-e   mētag-a   **pōjēn-e**
    lunch   food   grandmother-GEN house-OBL   send.PRS-2PL
    'She said {to her companions}: "Send some food for dinner and lunch to the old woman's house!"'   [Shi YN 1.170]

(62) *ballok=ē* **gwašt**
     grandmother=PC.3SG say.PST
     'He said {to the old woman}: ...' [Shi YN 1.135]

(63) *ǰanek-ā bačak-ā* **gwašt**
     girl-OBL boy-OBL say.PST
     'The boy said {to the girl}: ...'[17] [Shi YN 1.47]

(64) *kolpa dar ā če tambū-a*
     female_servant out come.PST.3SG from tent-OBL
     gō ōtī bānok=ē **gwaš**
     with own lady=PC.3SG say.PST
     'The servant came out of the tent and said {to her lady}: ...' [Shi YN 1.24]

In (65)–(66), 'say' is followed by quoted speech and an infinitive construction, respectively.

(65) **b-rō-∅** *čok-ān* **bo-goš-∅**
     IRR-go.PRS-IMP.2SG child-OBL.PL IRR-say.PRS-IMP.2SG
     hēč-a **ma-goš-an**
     anything-OBL PROH-say.PRS-3PL
     'Go [and] tell {the children} not to say anything (lit. [that] they should not say anything)!' [Ka GH 1.42; K&N 66]

(66) *čo ǰanēn drost-ā̊* **goš-ī**
     like woman all-OBL.PL say.PRS-3SG
     dast-ān pād-ān prēnč-ag-ā
     hand-OBL.PL foot.OBL.PL press.PRS-INF-OBL
     'Thus the woman told (lit. tells) {everyone} to press [her] hands and feet.'[18]
     [Shi YN 175f.]

(67) *mard ǰan-ā habar* **ko**
     man woman-OBL news do.PST
     ' The woman talked {to the man}

(68) *mard-ā ǰwāb* **na-dā**
     man-OBL answer NEG-give.PST
     [but] the man did not give an answer.' [Shi YN 1.215f.]

---

17 Only the context enables one to determine who is speaking in this example.
18 A Balochi bride is supposed to sit motionless for long hours; the other women and girls press her hands and feet to prevent them from becoming stiff.

The data suggest that the preverbal position is generalized for Addressees. One important factor surely is that the verb is usually followed by quoted speech; the "space" after the verb is thus already quite "full". This is the case in (61)–(64), where 'say' is directly followed by quoted speech. However, sometimes nothing follows, and the Addressee is still preverbal, as in (67) and (68).

An apparent counter-example is (69)–(71), where 'give' plus the curse could be interpreted as a speech verb ('to curse'), and has all arguments following the verb. However, the passage is more likely to be an instance of verb-fronting (see Section 6.2) expressing strong emphasis (roughly: "and he did curse him, he really did!").

(69) ***dāt-ag=ī***   *ham-ē*   *hākem-ā*   *zā*
    give.PST-PRF=PC3SG   EMPH-PROX   chief-OBL   curse
    'He gave {the chief} a curse (cursed the chief).'

(70) ***dāt-ag=ē***   *hākem-ā*
    give.PST-PRF=PC.3SG   chief-OBL
    He cursed {the chief}.

(71) ***dāt-ag=ē***   *hākem-ā*   ***dāt-ag=ē***
    give.PST-PRF=PC.3SG   chief-OBL   give.PST-PRF=PC.3SG
    He cursed {the chief}; he cursed [him].'   [Ko SE 1/1.83-86]

3.5 Final state

(72) illustrates both the postverbal and preverbal positions of a Final state:

(72) ***b-el-∅=ē***   ***bay-ān***   *šīr*
    IRR-leave.PRS-IMP.2SG=PC.3PL   become.PRS-3PL.SBJV   milk
    šīr   *a*   ***bay-ān***
    milk   VC   become.PRS-3PL.SBJV
    'Leave it so that it becomes {yoghurt}; it should become yoghurt.'[19]
                                                            [Ka GH 1.11; K&N 45]

More data would be needed to determine which position is preferred for such a Target. At any rate, the passage shows that the postverbal position is possible.

Note that *bay-* 'become' in the sense of 'turn into' patterns differently from its sense 'be' in a sentence such as in 'Be my friend!' In the latter use, the argument is in preverbal position in the data.

---

19 *šīr* is used for both 'milk' and dairy products such as yoghurt.

4 Other postverbal arguments

For a discussion of the position of Targets, it is relevant to check which other arguments (if any) can occur postverbally, and under which conditions. Also, the distinction between Targets and other arguments is not always clear-cut.

4.1 Adverbials

Somewhat related to Targets are **purpose-like** arguments and movements towards an abstract Goal, which are Goal-like in a metaphoric sense. They are indeed found in postverbal position (73), (74), cf. also the wedding in (8)–(9) and the hunt in (10)–(13). The postverbal position of purpose arguments may also be motivated by their clause-like features, as illustrated by the infinitive in (73), which is in the oblique case; it is thus parallel to datives used for "in order to" expressions in some languages.

(73) **šot-an**   ya   šap=ē   čār-ag-ā
     go.PST-3PL   one   night=IDV   look.PRS-INF-OBL
     'They went one evening {for visiting}.'   [Shi YN 1.174]

(74) gandīm-a   čol   **kod-ag-ā**   pa   hodā-ī   ēmīd-a
     wheat-OBL underground do.PST-PRF-3PL for   god-GEN hope-OBL
     '(In former times,) they put the seeds underground, {for the hope of God [that it would rain]}.'   [Ka GH 2.6]

(75)–(76) is somewhat between Goal and **location**;[20] it shows again the argument in postverbal position in the first instance, and preverbally directly afterwards. (78) could be seen as an instance of location or direct object; at any rate, the argument is extremely long.

(75) **bast-ag**=ī   gō   kahīr=īy-a
     bind.PST-PRF=PC.3SG   with   (tree)=IDV-OBL
     '... [and] he bound [it (the rope)] {to a Kahur tree}.
(76) gō   kahīr=ē   **bast-a**
     with   (tree)=IDV   bind.PST-PRF
     He bound [it] {to a Kahur tree}.'   [Ko SE 1/1.43f.]

---

20 gō(n) 'with' is used for concomitance (64), (81), (84) and instruments (82) and can also be employed as an adverb, as in (80), which recalls German verbs with *mit* such as *mitkommen*, *mitgeben* (lit. come with, give with). The latter in turn is parallel to verbs meaning 'put', see Section 3.3.3 and fn. 16.

(77) čār   andām=õ   mana   **bast-ag-ã**
     four   bodypart=PC.1SG I.OBJ   bind.PST-PRF-3PL
     ham-ā   bon-ā
     EMPH-DIST base-OBL
     'My four feet are tied {right at the base (at the ankle)}.'   [Ko GO 1.72]

(78) mošk-a   **gadruše**   ǰāh=ē   gōžd   o   ǰāh=ē   sād
     mouse-OBL gnaw.PST   place=IDV   flesh   and   place=IDV   rope
     o   ǰāh=ē   gōžd   o   ǰāh=ē   sād
     and   place=IDV   flesh   and   place=IDV   rope
     'The mouse gnawed, {here (lit. one place)} the meat, {there} the rope,
     {here} the meat, there the rope.'   [Ko GO 1.90]

(79) and 'one night' in (73) are instances of location in **time** expressed postverbally, although such expressions prefer the clause-initial position, e.g., 'next year' in (26) and 'the third day' in (38). In (9), 'the next day' is in non-initial preverbal position.

(79) pas-ān   rāh   a   **d-ēn**   sabāh-ā
     goat-OBL.PL   way   VC   give.PRS-1PL   morning-OBL
     'In the morning, we send the sheep {to the pasture}.'   [Ka GH 1.3]

Both **instruments and concomitance** arguments are commonly found in postverbal position. The preverbal position also occurs but seems less frequent.

(80) taw   ham-ē   labz-ā   **bo-gwaš-∅**   gō
     you.SG   EMPH-PROX   word-OBL   IRR-say.PRS-IMP.2SG   with
     'Repeat this word {after me (lit. with [me])}.'   [Shi YN 1.247]

(81) o   inčolo   inčolo   sōrū=ē   **dāt-ag-an**   gō   pōg-ān
     and   little   little   corn=PC.3PL   give.PST-PRF-3PL   with   chaff-OBL.PL
     'And they gave them very little corn, [which was] {with chaff}.'
                                                                  [Ko SE1/2.9]

(82) **zort-a**   ōtī   dast   **borret-ag-ã**   gō   kārč-a
     seize.PST-PRF   self.GEN   hand   cut.PST-PRF-3PL   with   knife-OBL
     'He seized [and] cut his own hands {with a knife}.'   [Ko SE 1/1.94]

(83) šīr   a   **day-ēn=ē**   gōṭok=ē   gōṭok=ē
     milk   VC   give.PRS-1PL=PC.3PL   rest=IDV   rest=IDV
     'We give them (our neighbors) milk; a little, a little.

(84) čok-olok   a   **kã**   gō   ōtī   ǰām-ok-ā
     child-DIM   VC   come.PRS.3PL   with   own   jar-DIM-OBL.PL
     The small children are coming {with their small jars},

(85) gōṭok=ē   gōṭok=ē   šīr   a   **dey-ēn**
     rest=IDV   rest=IDV   milk   VC   give.PRS-1PL
     we give [them] a little milk each.'   [Ka GH 1.116–118]

The postverbal position is also possible for **ablative** expressions, cf. (64) and (86)–(88). The preverbal position seems more common, though; it is illustrated in (3), (45), (51).

(86) **ǰest-en**   **tatk-en**   dar   **ātk-en**   če   dabestān-ā
     jump.PST-1PL   run.PST-1PL   out   come.PST-1PL   from   school-OBL
     'We came {out of school} running.'   [Ba RB 1.8; K&N 60]

(87) dar   ā   če   taht-ē   čēra
     out   come.PST.3SG   from   bed-GEN   under
     'He came out {from under the bed}.'   [Shi YN 1.232]

(88) māhī   ēr   **gept-ant**   a   lōg-ē   sarā
     fish   down   take.PST-3PL   from   house-GEN   on
     o   **wārt-ant=ḗ**
     and   eat.PST-3PL=PC.1PL
     'We took the fish down {from up in the house} and ate them.'   [Ka AJ 1.19f.]

4.1.1 Discussion

The various adverbials discussed in this section are quite frequently found in postverbal position, apparently without any pragmatic trigger being necessary for some types of adverbials. Information structure seems to play a role for some of them similar to that seen for Targets in Section 3.

Some of the adverbials are Goal-like such as the purpose arguments at the beginning of this section, but others are not. Particularly noteworthy are the instances of 'come' (which rarely occurs with a Goal, see Section 3.1.3) with "ablative" expressions; their position after the verb is the opposite of what one might expect by the iconic principle mentioned in Section 1.2.

4.2 Direct objects

Direct objects in postverbal position are surprisingly frequent.

(89) **goš-ī**   mã   har-ǰāgāh   **be-gend-ā̃**   kahīr-ā
     say.PRS-3SG   I   every-place   IRR-see.PRS-1SG   PN-OBL
     'He said, "Wherever I see {Kahir},

(90) kahīr-ā **koš-ā̃**
PN-OBL kill.PRS-1SG
haurokān-ā **zīr-ō̃** **k-āy-ō̃**
PN-OBL seize.PRS-1SG K.IND-come.PRS-1SG
I will kill {Kahir} [and] bring back {Haurokan}!'" [Ba SB 2.53; K&N 55]

(91) ešī-ā **zort** mohr-ēn=ē
PROX-OBL seize.PST strong-ATTR=IDV
'He (the fox) took {a strong (rope)}.' [Ko GO 1.29]

(92) **bort-ag-ant**=ē mē gohār
carry.PST-PRF-3PL=PC.3PL we.GEN sister
'They carried [away] {our sisters}.' [Ko SE 1/2.1]

(93) nī **by-ār-∅** dēz-a
now IRR-bring.PRS-IMP.2SG bowl-OBL
'... then bring {a pot}.

(94) manǰal-a **by-ār-∅**
pot-OBL IRR-bring.PRS-IMP.2SG
Bring {the pot}.' [Ko SE 2.53f.]

It seems to me that the postverbal position of direct objects is often triggered by pragmatic factors, but these are not always obvious. It might be so that, as in (69)–(71), where the indirect object is in postverbal position by fronting of the verb for emphasis, some postverbal direct objects could be explained as verb fronting as well. Possible instances are (6), (7), (24), (39), (52), (78), (82) and (92). Direct objects following a non-phrase-initial verb include (55) and 'a little' in (83), which might be accounted for by the direct object being new information. The passage (93)–(94) recalls passages with a Goal or a Recipient which is postverbal in the first clause and preverbal in the next sentence.

Note also that 'bring' and 'carry' (both infrequent with Goals, see Sections 3.1.3, 3.2.3) are commonly used with a direct object.

5 Summary

5.1 Hierarchy of postverbal-ness of Targets

While more research needs to be done to determine the relative frequency of post- and preverbal position of Targets, this survey suggests that the Targets listed in (1) are found in the following positions in the Southern Balochi data:

(95) Position of Targets in Southern Balochi according to this study

	preverbal	postverbal
a. Goal of verbs of motion	+	+ default
b. Goal of caused motion	+	+ default for 'carry' etc., not default for 'put'
c. Recipient of verbs of transfer	+	+ preferred: mirror position with direct object
d. Beneficiary		
e. Addressee of speech verbs	+	–
f. Final state	+ (1 ex.)	+ (1 example)

These results show interesting agreements and differences to those obtained by Jahani (2018) for several varieties of Balochi. In her study, postverbal Goals of motion verbs are found in all varieties of spoken Balochi, and this is also by far the preferred position in the two varieties which she selected for a closer study.

There is a major difference for the other Targets, though. Koroshi shows a significant preference for postverbal Recipients while these do occur, albeit only in a small minority of cases, in Southern Balochi. Koroshi also has postverbal Addressees and Final states, neither of which occur in Jahani's Southern Balochi data, while Western Balochi has postverbal Addressees. Jahani (2018) further shows that, just as standard Persian, written Southern Balochi is strongly verb-final; it does not show any postverbal Targets.

(96) Postverbal Targets (and their frequency) according to Jahani (2018)

	Koroshi		Western Balochi		Southern Balochi (Iran & Pakistan)	
			Sistan	Turkmenistan	oral	written
a. Goal of verbs of motion	+	92%	+	+	+ (90%)	–
c. Recipient	+	73%	+	+	+ (5%)	–
e. Addressee	+	8%	+	+	–	–
f. Final state	+	50%	–	–	–	–

Concerning Southern Balochi, Jahani's results largely agree with my own, but her very low figure for postverbal Recipients is unexpected in the light of my data. As for Final states, my data suggest that the postverbal position is possible in spoken Southern Balochi, and it might be due to them being overall rather rare that there are no instances in Jahani's data.[21]

5.2 Default vs. marked positions

5.2.1 Types of Targets

Concerning **Goals of verbs of motion** (Section 3.1), both this study and Jahani's 2018 results indicate that the postverbal position is the default one and does not need a specific motivation, irrespective of the length of the Goal and of it being an NP, a PP or an adverb such as 'here'. This is particularly clear for non-human Goals of motion, which could be called prototypical Targets, but also applies to [+human] Goals of motion.

It is rather the preverbal position that seems to occur only under specific conditions. As seen in the pairs of examples cited in Section 3.1.3, being mentioned (immediately) before is one motivating factor. For the Western Balochi of Sistan, Barjasteh Delforooz (2010: 61) comments on preverbal Goals:[22] "the expression is now given information, it has a function similar to that of tail-head-linkage" (for which see Sections 2.4 and 6.2).[23]

While this is surely correct, it does not imply that the postverbal Goal in the sentence preceding the preverbal Goal needs to be the first mention of that Goal, i.e., a Goal does not need to be new information to occur in postverbal position. Conversely, the fact of being mentioned earlier does not explain all instances of preverbal Goals. Perhaps, fixed expressions or literary style account for a stricter verb-final structure in examples such as (15) and (17).

One might think that **Goals of verbs of caused motion** behave like those of verbs of motion, as both imply a movement towards a certain point. This is not the case in my data: No clear preference stands out for the position of Goals of verbs of caused motion (Section 3.2). An exception to this are pairs of sentences such as those mentioned above for verbs of motion, where there is first a Goal in

---

[21] Note that she has four instances of Final states in Koroshi altogether, and she does not say how many (if any) she found in preverbal position in Southern Balochi, i.e. the total number of Final states in her data is not clear.

[22] His two examples of preverbal Goals are instances of the type (15)–(16), (18)–(19), i.e. they follow a sentence with postverbal Goal.

[23] He also says: The postverbal position of a goal serves "either to highlight an expression or to move the events along" (Barjasteh Delforooz 2010: 60), but this feature seems to be difficult to assess.

postverbal position and the second one occurs preverbally. However, such instances are markedly less frequent for caused motion. Indeed, the impression from my data that there is no clear preference for Goals of caused motion seems to come from 'put' being included in this group, for which the preverbal position is common or perhaps even default; excluding these, the postverbal position seems to be the preferred one. Barjasteh Delforooz (2010: 64) likewise seems to imply that the default position for Goals of caused motion in Sistani Balochi is postverbal.

Nevertheless, very differently from Goals of motion verbs, Goals of caused motion verbs other than those meaning 'put' are rather rare; speakers tend to say 'go to [place] and give [object] to [person]' or 'bring [object] and give to [person]' (see Section 6.1). In such a situation, the **Recipient** of 'give' typically is in preverbal position (Section 3.3). As **Beneficiaries** are less commonly part of such a chain, there is one motivation less for them to occur in preverbal position.

When not occurring in a series of verbs, Recipients and Beneficiaries are placed after the verb, particularly when they are new information. Stronger than this, however, is the tendency of having Recipients and direct objects in "mirror position" on opposite sides of the verb. This is in interesting contrast with Sistani Balochi, for which Barjasteh Delforooz (2010: 64) notes Recipients and direct objects both being in preverbal position, as also shown by his examples. He has an example of a postverbal Beneficiary, though.

This being so, it is not surprising that **Addressees** (Section 3.4) are in preverbal position, as the quoted speech following the verb functions as its direct object; and this position seems to be generalized also to the cases where there is no quoted speech in the text. It has to be kept in mind that overt Addressees are rare, as they are usually retrievable from the context.

There is one passage in the data with a Target of **Final state**, first in postverbal, then in preverbal position (Section 3.5), but of course, this is not sufficient material to determine the default position. Barjasteh Delforooz (2010: 61) says that the complement of 'become' is in preverbal position. His examples are all with adjectives (such as 'he became mad'). It seems possible to me that such cases are treated differently from those 'turn into X' (e.g., milk becomes yoghurt; someone is turned into a frog).

5.2.2 Other parameters

While the Goal of a motion verb is overwhelmingly [–human], Recipients, Beneficiaries and Addressees are typically **human**, and Addressees always so in my data. Needless to say, other living beings or abstract entities are potential Recipients and Beneficiaries, too ('give water to the flowers / give money to a worthy cause'), as are personified non-human Addressees.

Unlike Persian and Kurdish, where Goals show different **morphology** depending on their position,[24] Goals are marked identically in any position in Balochi.[25] For [–human] ones, the oblique case is sufficient for the marking of a Target (see Section 2.4) while [+human] ones show patterns containing oblique and genitive markers (cf. also Section 6.4).

The fact that [+human] Targets are often used with adpositions would contribute to the **length ("weight")** of the argument. It is a question for further research whether length of the argument plays a role for its placement, e.g., whether a Target with an adposition is more frequent in postverbal position than a Target without. Considering examples of very long preverbal Targets of 'put', it seems that, if length plays a role, other tendencies are liable to override this factor.

In this context, it is worth noting that nouns are frequently preceded by a **demonstrative** pronoun in Balochi (*ē, ā*, often with an originally emphatic element: *ham-ē, ham-ā*). Goals of motion and of caused motion containing a demonstrative are most often in postverbal position (4, 22, 27, 32, 33); the local argument 'at the base' in (77) is likewise postverbal. The preverbal position is much less common, but does occur (17, 19, 37, 44). Interestingly, the instances of Recipient with demonstrative are all preverbal (55, 57, 58).

5.3 Non-Target postverbal elements

As pointed out in Sections 1.1–1.3, several Iranian languages show Targets (particularly Goals of verbs of motion) in postverbal position, either frequently or, in some configurations, even as a rule. The question thus arises whether the postverbal position of elements other than Targets could be derived from this situation, i.e., whether the postverbal position could have been generalized as an option for non-Target arguments in one way or the other.

---

24 Cf. standard Persian *man be xāne mīravam* (I to house go.PRS.1SG) 'I go home' vs. colloquial *miram xune* (go.PRS.1SG house), where the preverbal Goal shows a preposition while the postverbal one does not (colloquial *xune miram* is likewise possible).
25 An exception to this is Koroshi, which, like colloquial Persian, shows postverbal Targets without oblique and/or adpositional marking. (The note about unmarked postverbal Targets in Nourzaei and Jügel 2018: 157 has to be read in this sense.)

The first candidate would be purpose-like arguments, which may be seen as Targets in a metaphorical sense.

Non-directional locations could be seen as an extension of the locative-like nature of non-human Goals of motion. Among these are arguments of concomitance and "ablative" expressions, which are found in postverbal position more often than one might expect. The postverbal position of "ablative" expressions (see Section 4.2) is particularly remarkable because it runs counter the iconicity principle mentioned in Section 1.2. If 'come' regularly occurred with a Goal (in postverbal position), one might think that the postverbal position was generalized for any argument of this verb, but the fact that it very rarely takes a Goal makes this assumption difficult.

As for the postverbal position of direct objects, which is not infrequent (Section 4.2), it is not quite clear whether it could be seen as an extension of the postverbal position of Recipients or rather of that of non-directional arguments. In both cases, the use of the oblique case for all these elements (see Section 2.4) might have played a role. The parallel treatment of Recipients and direct objects could also have been generalized from complex predicates, where an object could be seen as indirect object (Recipient or Beneficiary) if the non-verbal part of the complex predicate is seen as the direct object, or, if the non-verbal part of the predicate is incorporated, as direct object.

As a working hypothesis, I suggest a multi-dimensional cline showing the generalization of the postverbal position as an option in Balochi (97), starting with non-human Goals of motion as the most typical postverbal Target:

(97) Multi-dimensional cline of postverbal arguments in Southern Balochi

non-human Goal of motion → human Goal of motion → Recipient, Beneficiary

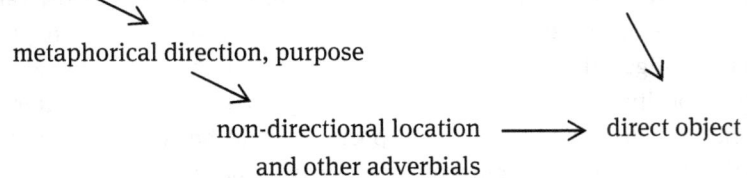

metaphorical direction, purpose

non-directional location ⟶ direct object
and other adverbials

6 Some other results of this study

As a by-product of the present study, the following elements suggest themselves as elements of Balochi grammar.

6.1 'come', 'bring' and serial verbs

The look at Targets undertaken in this paper shows that the conventional English translations of some verbs are to some extent misleading as they suggest the presence of arguments which they do not regularly take in Balochi: While the extremely frequent verb 'go' is commonly used with Goals, this is not at all the case for **'come'**, which seems to already imply a Goal 'here' (be this the place of the speaker or the 'here' of a specific point of a narrative, cf. (46) and (90). If the movement is not towards 'here', it is 'go' which tends to be used ('go there', 'go to someone'). Conversely, 'come' is often used with the origin of the movement; its sense is thus rather 'come here from X', and it is often used in this way in combination with other verbs.

The same is true for **'bring'**, which rather means 'bring here'. Instances such as (34) 'I brought you to my king' do occur, but more typically one finds 'X brings Y (scil. here) and gives him/her to the king', as in (60).

Even more frequently, the bringing is decomposed, as it were, into a chain of verbs describing the various phases of the action, thus 'go to (place or person X) [and] (object Y) give' (thus without a verb 'bring'), or, in a more elaborate form, '... comes from..., goes in direction of..., goes, goes, goes to (Target), carries Y, brings and gives'. The idea of 'bring' is thus not that of transporting an object to a place, but to carry it somewhere in order to give it to someone (cf. Nourzaei and Jügel 2018: 160). Similarly, instead of 'send', speakers commonly use a series of verbs such as the examples just cited, or simply 'give'.

This highlights an important point: While 'come' and 'bring' as such are very common, they typically occur in a chain of **serial verbs**. Syntactically, this has the effect of assigning the Target of the overall action to the verb of motion or the verb of transfer ('give'). Examples include 'carry' in (44) and 'bring' in (33), (38), (41), (44); further examples are found in Section 3.3.[26]

These patterns add to the frequency of 'give' occurring with a Target. One quite common type of pattern featuring 'give' is a series of the type 's/he went – to X ['s place] – object – gave'. They could be interpreted as either "s/he went to X's place [and] gave [him/her] the object", or as "s/he went [and] gave X the object". In my view, X is more likely to be the Recipient of 'give' rather than the Goal of 'go'. One might also argue that in such a series, 'go' is used as a mere vector verb and that the sentence expresses one event. However, as there is an actual movement involved, this is difficult to prove.

---

[26] Both 'carry and 'bring' are frequently used with direct objects (see Section 4.2).

6.2 Information structure

While matters are rather clear as to the default position of Goals, various other parameters play a role in governing the placement of other arguments. This in turn has the effect that it is sometimes difficult to decide which factor is present in a given example, with no tendency being entirely systematic.

As seen throughout this article, there is an overall tendency of new information being placed after the verb and given information in front, particularly in the rather common case of repetition of material from the preceding sentence (Tail-Head-Linkage), where a postverbal argument from the first sentence is often repeated in preverbal position, as in (10)–(13), (15)–(16), (18)–(19), (20)–(21), (23)–(24), (28)–(29), (93)–(94).

Also not rare is the fronting of the verb, including the second sentence of THL passages. Indeed, it seems to me that some of the examples cited in previous studies on the postverbal position of certain arguments in Iranian in fact show fronting of the verb. While the issue has not been investigated yet, I think that verb fronting either occurs in pragmatic situations such as 'I hereby declare...'[27] or else when there is a particular emphasis, as is clearly the case in (69)–(71) and (82), both referring to a rather dramatic situation. Note that in (51)–(52), the direct object *kāgad* 'letter' is preverbal in the first instance and postverbal in the second one, thus contrary to the pattern just noted; (52) thus seems to be an instance of verb fronting. In (51), the preverbal position of the direct object could be caused by the preference for the "mirror position" to the Recipient (see Section 3.3). In other cases, the reason for the fronting of the verb (if it is one, and not another phenomenon) is not clear to me.

Likewise to be kept in mind is the fact that any argument may be put in postverbal position for emphasis. This includes the subject in (98), where the intonation puts a strong emphasis on *šēr*. In (99), the genitive-marked dependent (usually found in front of the head noun in Balochi, see Section 2.1) is even separated from its head noun; this may be an example of an afterthought, as might be the case for the knife in (82).

(98) *goṛā   yak   rōč=ē*
     thus   one   day=IDV
     ham-ē   molk-e   bādšah   at   šēr
     EMPH-PROX   country-GEN   king   COP.PST.3SG   lion
     'Thus, one day, it was {the lion} who was this country's king.'

---

[27] One such example is the Old Persian phrase (occurring in numerous places in the Achaemenid inscriptions) *θātiy* (name of the king) ..., which has sometimes been cited as evidence for postverbal arguments in Old Iranian.

(99) šēr   bādšāh **en**   hokūmat   **en**   molk-ē
    lion   king   COP.PRS.3SG   government   COP.PRS.3SG   country-GEN
    [The animals were talking:] "The lion is the king. He is the governor, {of the country}."   [Ko GO 1.3f.]

6.3 Subordination

Subordinate clauses are often marked as such by the subordinator *ke*, which is phonologically enclitic to the preceding word. This means that a subordinate placed before the main clause and consisting only of the Target and the verb such as (19) *hamā molkā ke šo* 'as he went to that country...' ((21) is an example of the same type) needs to have the Target in preverbal position, because both †*ke šo hamā molkā* and †*šo ke hamā molkā* are ungrammatical.

However, this does not mean that a Target is necessarily preverbal in a subordinate clause. One option would be *hamā molkā šo* or *šo hamā molkā*, where the subordinate is marked by intonation only. This is frequently the case in a sequence with THL (see Sections 2.4, 6.2) as in (24), (29) and (52), but also occurs without repetition of material from the preceding sentence (cf. Korn 2017). Needless to say, the problem does not pose itself when additional elements are present because *ke* can then be enclitic to these.

6.4 Local reference to humans

In the course of the present study, it turned out that reference to a person's location is achieved by the paraphrases mentioned in Section 2.4 not only sporadically or as an option, but systematically, i.e., contrary to inanimates, the oblique alone cannot be used in Southern Balochi. As in many other languages, the location is conceived as 'at X's place'. The combination of the endings of the genitive and the oblique, entirely parallel to English *at my uncle's* [scil. *place*], is occasionally used (34). Much more frequently, however, adpositions are used, e.g., the postposition *gwar-ā* (side-OBL, lit. 'at the side of'), as in (20)–(21). This is somewhat similar to Azeri Turkic, where the use of the locative suffix is blocked on humans, and one employs the associative suffix *-gil*, e.g., *Murad-gil-də* 'at Murad's place' (lit. at-with-Murad).

Likewise frequent are expressions such as 'at X's house', as in (22), (46), (61); this recalls French, where local reference to a person is obligatorily with the preposition *chez* (e.g., *chez mon oncle*), which goes back to Latin *casa* 'house'.

7 Conclusion

While Balochi surely is an OV language, the frequency with which material occurs after the verb suggests that some arguments might be regularly found in this position, and that others may be placed there under certain syntactic or pragmatic conditions.

The former is the case for Goals of motion, perhaps also for Goals of caused motion (to the extent that these take a Target at all) and for certain types of adverbials, while the latter may apply to Recipients / Beneficiaries, other adverbials and direct objects. The fact that direct objects in the postverbal position occur at all, and are in fact not rare, is a noteworthy difference to Kurdish, where "[d]irect objects may be fronted for pragmatic purposes, but are only exceedingly rarely positioned after the predicate" (Haig 2015: 408).

For the Balochi of Sistan, Barjasteh Delforooz (2010: 61–63) assumes influence of colloquial Persian to account for the postverbal position of Goals of motion verbs, and Jahani (2018) refers to colloquial Persian for postverbal elements in general. Note, however, that Jahani (2018) includes data from Pakistan, where Balochi dialects have not been under particular Persian influence in recent centuries. Furthermore, assuming Persian influence only transports the problem to the next level, i.e., where then Persian got its postverbal elements from.

In the light of how common postverbal Goals are across the region, it seems to me that one could draw quite the opposite conclusion: If there is a preference for the postverbal position for certain types of Targets in a number of Iranian languages from various regions, this could indicate either a shared inheritance of this feature, and/or confirm Haig's conclusion from a survey of a large number of languages "that OV languages generally seem to permit a certain amount of "leakage" rightwards of the predicate, and that Goals are among the most frequently post-posed constituents. This fact suggests that they are less strictly linearized, and hence inherently more susceptible to contact influence" (Haig 2015: 414).

Concerning the latter point, it needs to be kept in mind that Balochi is outside the Iranian-Semitic contact zone which has been assumed to have caused the occurrence of postverbal Targets in Kurdish and Aramaic (see Section 1.3). Shared inheritance thus remains an option to investigate (see also Jügel, this volume).

Acknowledgments: I am very grateful to the editors, Thomas Jügel and Hiwa Asadpour, for their invitation to contribute to this volume, as well as for important feedback on an earlier version of this paper, which is reflected in a number of points discussed above. My thanks also go to Murad Suleymanov for comments and to an anonymous reviewer for checking the examples.

Abbreviations

Place names of the recordings

Ba[hukalat], Ka[rewan], Ko[narak], Shi[rgwaz];
these are followed by the initials of the speaker and the number of the text and sentence.

Published examples

K&N = Korn & Nourzaei (2019); the text Ko GO 1 is published in Korn & Nourzaei (forthcoming).

Glosses

ATTR	attributive suffix (-ēn)
CAUS	causative
COP	copula
DIM	diminutive
DIST	distal demonstrative
EL	elative form of adjective (-tir)
EMPH	emphatic marker ((ha)m-)
GEN	genitive
IDV	individuation marker (=ē and variants)
IMP	imperative
IRR	irrealis prefix (be- and variants)
k.IND	prefix k- of the indicative present of some verbs
OBJ	object
OBL	oblique
PC	pronominal clitic
PL	plural
PN	name
PRF	perfect
PROX	proximal demonstrative
PRS	present
PST	past
PTCP	participle
REFL	reflexive
SG	singular
SUB	subordinator (ke)
VC	verbal clitic (TAM marker, a)

References

Asadpour, Hiwa. this volume. Word order in Mukri Kurdish – the case of incorporated targets.
Barjasteh Delforooz, Behrooz. 2010. *Discourse features in Balochi of Sistan (oral narratives)*. Uppsala: Uppsala Universitet. http://uu.divaportal.org/smash/record.jsf?pid=diva2: 345413.
Bashir, Elena. 2008. Some transitional features of Eastern Balochi: An areal and diachronic perspective. In Carina Jahani, Agnes Korn & Paul Titus (eds.), *The Baloch and others. Linguistic, historical and socio-political perspectives on pluralism in Balochistan*, 45–82. Wiesbaden: Reichert.
Boyajian, Vahe. 2003. Towards the interpretation of the term *balōč* in the *Šāhnāme*. In Carina Jahani & Agnes Korn (eds.), *The Baloch and their neighbours. Ethnic and linguistic contact in Balochistan in historical and modern times*, 313–320. Wiesbaden: Reichert.
Dryer, Matthew S. 2013. Relationship between the order of object and verb and the order of adposition and noun phrase. In Matthew S. Dryer & Martin Haspelmath (eds.), *The world atlas of language structures online*. Leipzig: Max Planck Institute for Evolutionary Anthropology. https://wals.info/chapter/95.
Haig, Geoffrey. 2015. Verb-goal (VG) word order in Kurdish and Neo-Aramaic: Typological and areal considerations. In Geoffrey Khan & Lidia Napiorkowska (eds.), *Neo-Aramaic and its linguistic context*, 407–425. Piscataway: Gorgias.
Haiman, John. 1983. Iconic and economic motivation. *Language* 59. 781–819.
Harris, Alice C. & Lyle Campbell. 1995. *Historical syntax in cross-linguistic perspective*. Cambridge: Cambridge University Press.
Jahani, Carina. 2018. Post-verbal arguments in Balochi. Paper presented at the Conference *Anatolia-The Caucasus-Iran: Ethnic and linguistic contacts*, Yerevan.
Jahani, Carina & Agnes Korn. 2009. Balochi. In Gernot Windfuhr (ed.), *The Iranian languages*, 634–692. London & New York: Routledge.
Jahani, Carina, Maryam Nourzaei & Behrooz Barjasteh Delforooz. 2012. Non-canonical subjects in Balochi. In Behrad Aghaei & Mohammad R. Ghanoonparvar (eds.), *Iranian languages and culture: Essays in honor of Gernot Ludwig Windfuhr*, 196–218. Costa Mesa: Mazda Publishers.
Jügel, Thomas. this volume. Word order variation in Middle Iranic: Persian, Parthian, Bactrian, and Sogdian.
Korn, Agnes. 2005a. Das Nominalsystem des Balochi, mitteliranisch betrachtet. In Günter Schweiger (ed.), *Indogermanica: Festschrift Gert Klingenschmitt. Indische, iranische und indogermanische Studien dem verehrten Jubilar dargebracht zu seinem fünfundsechzigsten Geburtstag*, 289–302. Taimering: VWT-Verlag.
Korn, Agnes. 2005b. *Towards a historical grammar of Balochi*. Wiesbaden: Reichert.
Korn, Agnes. 2006. Counting sheep and camels in Balochi. In Mixail N. Bogoljubov et al. (eds.), *Indoiranskoe jazykoznanie i tipologija jazykovyx situacij. Sbornik statej k 75-letiju professora A. L. Grjunberga (1930–1995)*, 201–212. St Petersburg: Nauka.
Korn, Agnes. 2008. A new locative case in Turkmenistan Balochi. *Iran and the Caucasus* 12. 83–99.
Korn, Agnes. 2016. A partial tree of Central Iranian: A new look at Iranian subphyla. *Indogermanische Forschungen* 121. 401–434.

Korn, Agnes. 2017. Subordonnées et leurs équivalents en baloutchi et bachkardi. https://halshs.archives-ouvertes.fr/halshs-01638071.
Korn, Agnes. 2019. Isoglosses and subdivisions of Iranian. *Journal of Historical Linguistics* 9. 239–281.
Korn, Agnes & Maryam Nourzaei. 2019. Notes on the speech of the Afro-Baloch of the southern coast of Iran. *Journal of the Royal Asiatic Society* 29. 623–657.
Korn, Agnes & Maryam Nourzaei. forthcoming. Getting rid of the lion: A Balochi fable. (for a Festschrift).
Nourzaei, Maryam. 2017. *Participant reference in three Balochi dialects: Male and female narrations of folktales and biographical tales*. Uppsala: Uppsala Universitet. http://uu.diva-portal.org/smash/record.jsf?pid=diva2%3A1069126&dswid=-6317.
Nourzaei, Maryam & Thomas Jügel. 2018. Ditransitive constructions in three Balochi dialects. In Agnes Korn & Andrej Malchukov (eds.), *Ditransitive constructions in a cross-linguistic perspective*, 147–163. Wiesbaden: Reichert.

Bruno Herin
Word order in contact and the expression of Target in Northern Domari

Abstract: The present paper investigates the morphosyntax of Targets in Northern Domari. Morphologically, it shows that Domari does not have a unified strategy to mark Targets. Syntactically, it appears the preverbal and postverbal positions are accessible to Target constituents. The data collected so far also suggest that there is no evidence that Target constituents exhibit a different syntactic behavior from non-Target constituents. Furthermore, it appears that contact is the most probable cause for the accessibility of both the preverbal and postverbal positions, as the outcome of the resolution of the conflicting orders of Arabic and Turkish.

Keywords: case marking; contact; Domari; Targets; word order

1 Introduction

Target is a semantic macro-role that subsumes the roles of goal of motion and caused-motion verbs, the recipient-like argument in ditransitive constructions, the addressee of say-verbs, the goal of inchoative verbs and beneficiaries (Asadpour, this volume), which are all linked to the semantics of 'endpoints', as stated by Haig (to appear). The present study investigates the morphosyntactic correlates of Target-marking in Northern Domari in order to see whether it patterns with some of the tendencies observed in neighboring West-Asian languages whereby Targets tend to occur rightward of the predicate. Domari also has the peculiarity of being a minority language in intense contact with Arabic and/or Turkish. Consequently, the present study also aims at documenting the effect of contact on word order.

Domari is an Indic language spoken by itinerant service-providing communities scattered across the Middle East. Dom clans in Northern Syria and Southern Turkey are now famous for their practice of informal dentistry (Bochi 2015) but claim jewellery as one of their previous crafts. The Doms of Beirut and Da-

---

Bruno Herin: Institut National des Langues et Civilisations Orientales, Rue des Grands Moulins 65, boîte 150, 75013 Paris, France, E-mail: bruno.herin@inalco.fr

mascus claim sieve-making as their main pre-modern occupation but now seem to have abandoned this practice in favor of a wide variety of petty itinerant trades. 'Dōm' is the only endonym used by these groups and is mostly unknown to outsiders, who refer to them with exonyms such as *nawar, qurbāṭ, qarač* and many others.

The data on which the present study draws were collected by the present author amongst Dom communities in Lebanon, Syria, Southern Turkey and more recently with Syrian Dom refugees in Europe. Initially, the corpus consisted for the most part of elicited data largely based on the questionnaire developed by Matras et al. (2001) as a tool to investigate and map Romani dialects. Three varieties of Domari were covered using this questionnaire: Aleppo (2010), Beirut (2011) and Sarāqib in Northern Syria (2011). Although the adaptation of this questionnaire to Domari posed a series of challenges, it allowed to cover much of the phonology and morphology of these varieties. Subsequent fieldworks focused on collecting less controlled data such as narratives, conversations and also elicited data using non-translational techniques such as the Pear Story (Chafe 1980), the Story Builder (Sardinha 2011), the Totem Field Storyboards[1] and other picture books (O'Shannessy 2004).

Domari is a contact language, which means that all its fluent speakers exhibit balanced bilingualism between Domari and the majority language, which can be either Arabic (in Syria, Lebanon, Jordan and Palestine) or Turkish (in Turkey). Because the use of Domari is limited to informal domains and in-group communication, Arabic and Turkish provide the "socio-cultural roofing"[2] in their respective areas. Domari is mainly restricted to oral communication, although literate speakers may write it in Latin or Arabic scripts in informal electronic exchanges.

From the point of view of Domari dialectology, two dialectal clusters have been identified: Southern Domari spoken in Palestine and Jordan (Matras 2012), and Northern Domari spoken in Lebanon, Syria and Turkey (Herin 2014, 2016). Northern Domari can be further divided into two sub-dialects: the Beirut-Damascus dialect, and the dialect of the clans located in Northern Syria (and Lebanon) and Southern Turkey (Figure 1).

---

[1] The Totem Field Storyboards is a collection of storyboards from various authors available at the time of writing at http://totemfieldstoryboards.org.
[2] The term is borrowed from Cerruti and Regis (2020).

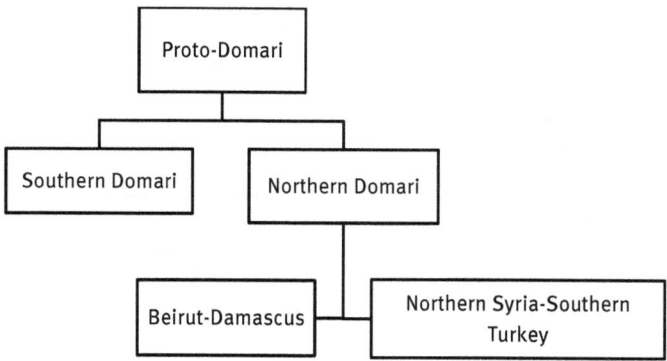

Figure 1: The dialects of Domari.

2 Case marking in Domari

Before breaking down the morphosyntactic correlates of Target-marking, a brief summary of case marking in Domari is needed. Domari is one of those languages which exhibit case-stacking, that is "the obligatory affixation of certain case markers to already case-inflected bases" (Iggesen 2013). Arkadiev (forthcoming) distinguishes three types of case-stacking: derivational case-stacking, case compounding and multiple case marking. Domari corresponds closely to the derivational type in that "one case form serves as a base for the formation of other cases" (Arkadiev forthcoming). This pattern is found in many Indic languages (Masica 1993). Similar to what is observed in Romani, Matras (2009, 2012) identifies three layers in Domari, exemplified in (1).

(1) z   dōm   -an   -ki
     from   Dom   -OBL.PL   -ABL
     Layer III X   Layer I   Layer II
     'from the Doms'

Layer I in Romani and Southern Domari is usually called the oblique. It differentially marks the object and is the base to which Layer II markers attach. In these varieties, oblique marking is gender sensitive in the singular. Northern Domari lost gender distinction and reduced exponency, reassigning the masculine oblique to a general singular differential object marker and the feminine singular oblique to a general oblique used in genitive constructions and to mediate between the base and Layer II markers. In addition to this, Northern Domari also exhibits the archaic marker -*ən*, reflex of an Old Indic instrumental that was re-functionalized as a differential subject marker, as summarized in Table 1.

Table 1: Layer I markers in Romani, Southern Domari and Northern Domari.

	Romani	Southern Domari	Northern Domari		
OBL.SG.M	-*es*	-*as*	-*as*	ACC (DOM)	< GEN.M.SG -*asya*
OBL.SG.F	-*(y)a*	-*a*	-*a*	OBL.SG	< GEN.F.SG -*āyaḥ*
OBL.PL	-*en*	-*an*	-*an*	OBL.PL	< GEN.PL -*ānām*
			-*ən*	NOM (DSM)	< INST.M.SG -*ena*

Table 2 shows the seven Layer II markers that attach to an oblique stem, here *pānī* 'water' to which the oblique marker -*a* suffixes.

Table 2: Layer II in Northern Domari: *pāny-a* 'water-OBL'.

Inessive	*pāny-a-ma*	'in/with the water'
Superessive	*pāny-a-ta*	'on the water'
Adessive	*pāny-a-ka*	'at the water'
Ablative	*pāny-a-ki*	'from/to the water'
Comitative	*pāny-a-sa*	'with the water'
Versative	*pāny-a-wa*	'towards the water'
Similative	*pāny-a-war*	'like water'

Domari also possesses a set of relational nouns that mostly express spatial relations as listed in Table 3.

Table 3: Relational nouns in Northern Domari.

pāny-a fatūn	'above the water'
pāny-a bna	'under the water'
pāny-a wagar	'in front of the water'
pāny-a pačī	'behind the water'
pāny-a manğ-a-ma	'inside the water'
pāny-a xr-a-ma	

pāny-a čanč-a-ta	'next to the water'
pāny-a čōrm-an-ta	'around the water'

Benefactivity is expressed with the morpheme *kēra* or its reduced form *kē*. It behaves phonologically like a relational noun, and morphosyntactically like a Layer II marker:

pāny-a kēra ~ pāny-a kē 'for the water'

In addition to this, Domari also has prepositions borrowed from previous and current contact languages. In Northern Domari, the most common is the Iranian derived preposition *z* 'from' and others, seemingly more recent, borrowed from Arabic such as *qabəl* 'before' and *baʕəd* 'after'.

3 Target-marking

Morphologically, Northern Domari makes use of four kinds of marking for Targets: ablative, benefactive and superessive as well as zero-marking. In a nutshell, ablative marking is used for the goal of motion verbs and the recipient of ditransitive verbs. Superessive marking is used for the addressee of 'say' verbs and in the speech of innovative speakers for the goal of motion verbs, replicating the Arabic pattern. Benefactive marking may surface in the speech of innovative speakers for the recipient of ditransitive verbs, replicating both Arabic *la* 'for' and the Turkish dative *-a*. Zero-marking occurs with goals of verbs of motion which denote topo-nouns (see Haspelmath 2019 for the cross-linguistic tendency for topo-nouns to be zero-marked, defined as a "set of nouns that denote concepts which are commonly used as spatial landmarks").

In Arabic, both SVO and VSO are unmarked and Targets occur rightward of the verb, although not in strict adjacency. Turkish is reported to be strongly verb-final and the unmarked order is described as follows: subject/modal adverbial – accusative-marked direct object – other adverbials – oblique object/non-case-marked direct object – verb (Göksel and Kerslake 2005: 338). Targets are not mentioned but they would fill the "other adverbial" slot, which occurs leftward of the verb.

3.1 Goal of verbs of motion

The goal argument of verbs of motion is encoded with the ablative marker -*ki* as in (2) or zero-marked (3).

(2) Sarāqib
    *har   dīs   **kam-ə-ki**   saʕat   ḥaft   rasame*
    every   day   work-OBL-ABL   hour   seven   arrive.IPFV.1SG
    'Every day I get to work at seven.'

(3) Sarāqib
    *har   dīs   **wyār**   ğām-a   ši*
    every   day   market   go.IPFV.1SG-PST   FOC
    'Every day I used to go to the market'

The linear order in (2) and (3) is T(arget) V(erb) but this order is quite marginal in the data collected in Syria and Lebanon in which the Target is most often placed to the right of the predicate as in (4) and (5).

(4) Aleppo
    *n-(h)ōre   ğās   **dāwat-ə-ki***
    NEG-become.IPFV.3SG   go.SBJV.2PL   wedding-OBL-ABL
    'You can't go to the wedding.'

(5) Saraqib
    *kā   pərtyəm-a   **krī***
    FUT   return.SBJV.1SG-PST   house
    'I wanted to go back home.'

Superessive marking is also possible in Beirut/Damascus Domari, but it replicates the Arabic pattern in which goals of motion verbs are marked with the preposition *ʕa(la)* whose core meaning is superessive 'on'. In (6), *mistašfē-ta* is reminiscent of Arabic *ʕa l-mistašfa* (on DEF-hospital). The order is also V(erb) T(arget), which is also the unmarked order in Arabic.

(6) Beirut
    *killma   ğām   **mistašf-ē-ta**   kā   gāl   lakar*
    whenever   go.SBJV.1SG   hospital-OBL-SUP   FUT   word   do.SBJV.3SG
    tō   bēn-ō-m   ištōr
    2SG   sister-SG-1SG   COP.2SG
    'Whenever I go to the hospital, he tells me "You are my sister."'

With pronominal Targets, all the recorded tokens occur rightward of the predicate as in (7).

(7) Sarāqib
    lāzim   *ağ*   *pərtyəm*   **wēšī**
    necessary today return.SBJV.1SG 3SG.ALL
    'I need to go back to it today.'

In Turkish Domari (here Hatay Domari), the unmarked syntax seems to be T(arget) V(erb), as suggested by the following sequence in (8) in which the speaker talks about previous patterns of mobility. In (b), the Target occurs after the predicate but the constituent *qlār-ən-ki* 'to the Bedouins' is prosodically salient in terms of pitch and intensity whereas in (c), the Target *šarq-ə-wa* 'towards the east' does not exhibit any acoustic prominence in the clause. In the current state of research, it is not clear whether focus marking stems from both postverbal position and prosody or simply from prosody, leaving the preverbal and the postverbal positions accessible to Targets, irrespective of their informational status. This state of affairs contradicts the Turkish pattern in which focused constituents "appear in the area preceding the predicate" and "backgrounded information follows the predicate and is never stressed" (Göksel and Kerslake 2005: 344).

(8)   Antioch
    a.   *čādr-ō-ma*   *ašt-a*   *ši*
         tent-SG-1PL   exist-PST   too
    b.   *ğān-a*   **qlār-ən-ki**
         go.IPFV.1PL-PST   Bedouin-OBL.PL-ABL
    c.   **šarq-ə-wa**   *ğān-a*
         east-OBL-VERS   go.IPFV.1PL-PST
         '(We) had a tent, (we) used to go to the Bedouins, eastward.'

3.2 Goal of caused-motion verbs

In all the collected tokens, the goal of caused-motion verbs always occurs after the predicate, as in (9) and (10).

(9) Sarāqib
     bāb-ō-s   *lğwāldōs*   **dād-ō-s-ka**
     father-SG-3SG   send.PFV.SBJ.3SG.OBJ.3SG   mother-SG-3SG-AD
     'His father sent him to his mother's.'

(10) Sarāqib

tōrdōm	kyā	**ṭāwl-ē-ta**
put.PFV.1SG	something	table-OBL-SUP

'(I) put something on the table.'

With pronominal Targets, the following orders are attested: VOT (11) and TOV (12). Although the effect of information structure on these word order variations is not straightforwardly assessable, the lack of clear-cut prosodic prominence of any constituent suggests that none of these orders is marked. Context is not available for (11) because it was elicited and in example (12), although it is extracted from a narrative, the Target constituent *abōr* 'for you' does not seem to bear any specific discursive role.

(11) Sarāqib

kā	nānəm	pānī	**abōs**
FUT	bring.SBJV.1SG	water	3SG.GEN

'I'll bring you water.'

(12) Sarāqib

abōr	čāġa	lǧwālda	rabbī	kar-əs
2SG.BEN	boy	send.PFV.3SG	raise	do.SBJV.2SG-3SG

'(God) sent you a boy for you to raise him.'

3.3 Recipient of ditransitive constructions: *dē-/tət-* 'give'

When the argument is a full NP, the most common marker is ablative *-ki* for the recipient-like argument in ditransitive constructions. As far as syntax is concerned elicited data yields a wide range of constituent orders: S(ubject) O(bject) V(erb) T(arget) as in (13), STVO as in (14), SVOT as in (15). It shows that not only both pre- and post-predicate orders are licensed but also that adjacency to the predicate is not required. The SVOT order, which is the Arabic order, is the most frequent in the data collected in Lebanon and Syria.

(13) Sarāqib

hā	qər	**kwāk-ki**	nə-dēre	kyāmōr
DEM	boy	someone-ABL	NEG-give.IPFV.3SG	anything

'This boy does not give anything to anyone.'

(14) Sarāqib

ē	qr-ənạ	kyāmōr	nə-ttạ	**kwāk-ki**
DEM.OBL	boy-NOM	anything	NEG-give.PFV.3SG	someone-ABL

'This boy didn't give anything to anyone.'

(15) Sarāqib
*šēx   dēre   qr-əs   **dādy-ə-ki***
elder   give.IPFV.3SG   boy-ACC   grandmother-OBL-ABL
'The old person gives the boy to the grandmother.'

Innovative speakers in both Syria/Lebanon and Turkey may replace ablative marking with benefactive marking, replicating the Arabic pattern (preposition *la* 'for') and the Turkish pattern (dative *-a*), respectively. In (16), the Target occurs leftward of the verb, which seems to replicate the Turkish order, albeit only partially because the canonical Turkish order is (S)OTV, whereas (16) displays TOV. This may be due to the focalization of the theme argument *sōwan* 'gold' as evidenced by the additive focus enclitic *ši*. The benefactive marker *kēra* is here realized *kēr*.

(16) Antioch
lāfty-ə   kēr   *sōwan   ši   dēnde*
girl-OBL   BEN   gold   too   give.IPFV.3PL
'(They) also give gold to the girl.'

Benefactive marking also appears in (17) but unlike (16) the recipient is placed postverbally, which agrees with the Arabic syntax.

(17) Beirut
pēndōs-s   ʕārḍ-a-ki   kā   bdēr-əs
lift.PFV.3SG-3SG   ground-OBL-ABL   FUT   give.SBJV.3SG-3SG
bēly-a   kēra
friend-OBL   BEN
'(He) lifted it (i.e., the pear) from the ground to give it to (his) friend.'

Pronominal recipients are indexed on the verb as objects (18):

(18) Antioch
*maʕāš   dēšt-ənd-**mān**-e*
pension   give.PROG-SBJ.3PL-OBJ.1PL-PRS
'They give us a pension.'

3.4 Addressee of 'say' verbs: *gāl kar-* 'say'

Addressees of 'say' verbs are normally marked with superessive case. That is the suffix *-ta* in the case of full NPs. In all the recorded tokens, the addressee argument occurs to the right of the predicate, as exemplified in (19) and (20):

(19) Beirut

	baʕd	fatr-āk	gāl	kra	**dād-a-ta**
	after	period-INDF	word	do.PFV.3SG	mother-OBL-SUP
	kā	ǧām	kəltyam		bāra
	FUT	go.SBJV.1SG	exit.SBJV.1SG		outside

'After a while she said to (her) mother "(I)'ll go out."'

(20) Aleppo
| | gārdōm | **ǧəʊr-ō-m-ta** | abōm | karər | qaḥwa |
| | say.PFV.1SG | woman-SG-1SG-SUP | 1SG.BEN | do.SBJV.3SG | coffee |

'(I) told my wife to make coffee for me.'

When the addressee is pronominal, it occurs equally to the left and to the right of the predicate (21):

(21) Sarāqib
| | gāl | kạ | **watōm** | ~ | **watōm** | gāl | kạ |
| | word | do.PFV.3SG | 1SG.SUP | | 1SG.SUP | word | do.PFV.3SG |

'(S)he told me.'

3.5 Complement of inchoative *hr-/hō-* 'become'

The verb *hr-/hō-* 'become' is used as an inchoative 'become' and also as a light verb of (mostly) intransitive predicates. The perfective stem is *hr-* and *hō-* is the imperfective stem. In all varieties of Domari, the complement of inchoative *hr-/hō-* occurs leftward and in strict adjacency (22).

(22) Sarāqib
| | hnū | **daktōr** | (h)ra |
| | DEM | doctor | become.PFV.3SG |

'(He) became a doctor.'

3.6 Non-Target oblique constituents

Non-Target constituents are attested both in postverbal (23) and preverbal position (24). Both examples were elicited so no contextual indication about their informational status is available. Prosodically, none of these constituents appear to be highlighted which suggest that they are not marked for discursive role.

(23) Sarāqib
 ǧib   nə-kōm   **kwāk-ə-sa**
 tongue   NEG-DO.PFV.1SG   someone-OBL-COM
 '(I) didn't speak with anyone.'

(24) Sarāqib
 kwāk-ə-sa   lagīš   nə-kēn
 someone-OBL-COM   quarrel   NEG-DO.PFV.1PL
 '(We) didn't quarrel with anyone.'

In terms of frequency, the Lebanese and Syrian corpora clearly exhibit a preference for the postverbal position, which expectedly patterns with the Arabic order as in (25).

(25) Sarāqib
 kā   wēštən   **ġēr**   **tāw-ə-t**
 FUT   sit.SBJV.1PL   other   place-OBL-SUP
 '(We)'ll sit somewhere else.'

In the corpus collected in Turkey, non-Target oblique constituents occur both to the left (26) and to the right of the predicate (27). In (26), *təkn-ə-m* 'in Turkey' is focused, which agrees with the Turkish pattern in which focused constituent are left-adjacent to the predicate or in their unmarked preverbal position (Göksel and Kerslake 2005: 344).

(26) Antioch
 āyrēnd   təkna   **təkn-ə-m**   māndēnd
 come.PFV.3PL Turkey   Turkey-OBL-IN   stay.PFV.3PL
 '(They) came to Turkey and stayed in Turkey.'

In (27), *sūriyyē-ma* 'in Syria' is part of the background, so its rightward position may also match the Turkish syntax where the postverbal position is accessible to background constituents (Göksel and Kerslake 2005: 345–346).

(27) Antioch
 dōm   nə-mānde   **sūriyyē-ma**   bitātan
 Dom   NEG-stay.PRF.3SG   Syria.OBL-IN   at_all
 'There are no Dom left in Syria at all'

4 Discussion and conclusion

The present study investigated whether Targets in Northern Domari follow the pattern displayed in many languages of Western Asia whereby Targets tend to occur in the postverbal position. Haig (to appear) provides three possible motivations for this tendency. The first one is an effect of syntactic iconicity in which spatial endpoints map syntactically onto clause final position. The second one is the avoidance of the so-called Final over Final Constraint-violation (Holmberg 2000) that states that OV (head final clauses) languages disfavor prepositional phrases (head initial phrases) in preverbal position. The third one is the effect of contact with languages that place Targets after the verb. From the data presented above, the following observations can be made:

(i) Targets behave similarly to non-Targets in that there is no evidence that Targets tend to occur postpredicatively more than non-Targets.
(ii) The Final over Final Constraint-violation does not seem to have much relevance in the case of Northern Domari because Targets are never realized as prepositional phrases.
(iii) The most relevant factor for explaining the occurrence of postpredicate constituents in Northern Domari seem to be contact.

As far as contact and word order is concerned, Northern Domari is an interesting case because it is strained by Arabic and Turkish[3], which have different word orders: Arabic is strongly VO and Turkish is strongly OV. In terms of contact, there are three situations. The most straightforward is the Beirut-Damascus dialect which is the most affected by Arabic, leaving the influence of Turkish marginal. The contact situation for the dialects of Northern Syria and Southern Turkey is more complex. Clans are spread across the Syrian-Turkish border and they used to travel back and forth when there were no borders or when borders were open, which suggests a symmetrical contact with both Arabic and Turkish. From the seventies onward, consultants claim that border crossing became much more difficult, forcing some clans to settle in Turkey, while others stayed

---

3 A reviewer notes that for the present argument to be convincing, reference to word order in the pre-contact stage should be made. First, diachrony is not accessible to investigation because Domari is mostly known from sources that are less than a century old. Second, there is probably no "pre-contact stage" because Domari speakers have been multilingual since the genesis of the community, given their itinerant profile of commercial nomads. The only suggestion that could reasonably be made is that before migration from Central India, the Indian forerunner of Domari patterned with the other languages of the area as far as word order is concerned, which is known for having evolved from a seemingly free order to a more rigid SOV order.

in Syria. When comparing recordings from Syrian and Turkish clans, it clearly appears that postverbal placement is much more pervasive in the Syrian data than in the Turkish data, although preverbal and postverbal orders are found both in Syria and Turkey. It was also shown that there is no firm evidence that information structure plays a decisive role in placing constituents preverbally or postverbally. The suggestion put forward here and supported by the data is that the syntactic conflict between the Arabic order and the Turkish order was resolved in Domari through licensing both orders, making the preverbal and postverbal positions accessible to these constituents. The prediction is that
(i) the preverbal position is being marginalized in the speech of the clans that remained in Syria, thus converging towards the Arabic pattern.
(ii) the postverbal position is being marginalized in Turkish Domari, except for topical constituents, in agreement with Turkish.

What remains to be investigated, in addition to refining the observations presented above with firm statistical evidence, is the impact of contact on word order typology as a whole, which includes the position of objects, in addition to Target and non-Target constituents.

Abbreviations

1, 2, 3	1st, 2nd, 3rd person
ABL	ablative
ACC	accusative
AD	adessive
ALL	allative
BEN	benefactive
COM	comitative
COP	copula
DEM	demonstrative
DOM	differential object marking
DSM	differential subject marker
F	feminine
FOC	focus
FUT	future
GEN	genitive
IN	inessive
INDF	indefinite
INST	instrumental
IPFV	imperfective
M	masculine
NEG	negation

NOM	nominative
OBJ	object
OBL	oblique
PFV	perfective
PL	plural
PROG	progressive
PRS	present
PST	past
SBJ	subject
SBJV	subjunctive
SG	singular
SUP	superessive
VERS	versative

References

Asadpour, Hiwa. this volume. Word order in Mukri Kurdish – the case of incorporated targets.

Arkadiev, Peter. forthcoming. Case-stacking. In *Wörterbücher zur Sprach- und Kommunikationswissenschaft (WSK) online*. Berlin & Boston: De Gruyter Mouton. https://www.academia.edu/36537075/Case_stacking.

Bochi, Giovanni. 2015. Exploring pluralism in oral health care: Dom informal dentists in Northern Lebanon. *Medical Anthropology Quarterly* 29(1). 80–96. https://doi.org/10.1111/maq.12066.

Cerruti, Massimo & Riccardo Regis. 2020. Partitive determiners in Piedmontese: A case of language variation and change in a contact setting. *Linguistics* 58(3). 651–677. https://doi.org/10.1515/ling-2020-0080.

Chafe, Wallace (ed.). 1980. *The Pear Stories: Cognitive, cultural, and linguistic aspects of narrative production*. Norwood: Ablex.

Göksel, Aslı & Celia Kerslake. 2005. *Turkish: A comprehensive grammar*. London: Routledge.

Haig, Geoffrey. to appear. Post-predicate constituents in Kurdish. In Yaron Matras, Ergin Öpengin & Geoffrey Haig (eds.), *Structural and typological variation in the dialects of Kurdish*. London: Palgrave Macmillan.

Haspelmath, Martin. 2019. Differential place marking and differential object marking. *STUF/Language Typology and Universals* 72(3). 313–334. https://doi.org/10.1515/stuf-2019-0013.

Herin, Bruno. 2014. The Northern dialects of Domari. *Zeitschrift der Deutschen Morgenländischen Gesellschaft* 164(2). 407–450. https://doi.org/10.13173/zeitdeutmorgese.164.2.0407.

Herin, Bruno. 2016. Elements of Domari Dialectology. *Mediterranean Language Review* 23. 33–73. https://doi.org/10.13173/medilangrevi.23.2016.0033.

Holmberg, Anders. 2000. Deriving OV order in Finnish. In Peter Svenonius (ed.), *The derivation of VO and OV*, vol. 31, 123–152. Amsterdam & Philadelphia: John Benjamins. https://doi.org/10.1075/la.31.06hol. https://benjamins.com/catalog/la.31.06hol (9 July, 2020).

Nicolaos Neocleous
The evolution of VO and OV alternation in Romeyka

Abstract: In this article, I investigate the evolution of VO and OV alternation in main and subordinate clauses in Romeyka, which is the only remaining variety of Asia Minor Greek that is still spoken in the area historically known as Asia Minor (present-day Anatolia, Turkey). The findings of this study indicate that (a) the pragmatically unmarked VO order in main clauses in Romeyka is the result of continuity from previous stages of Greek and (b) the pragmatically unmarked OV order in subordinate clauses in Romeyka is the result of external change due to contact with local Turkish.

Keywords: Asia Minor Greek; Anatolian Turkish; VO-to-OV change; SEM-(un)interpretable features; language contact

1 Introduction

In this article, I investigate the evolution of VO and OV alternation in Romeyka, a variety of Asia Minor Greek (henceforth AMG). I specifically (a) reconstruct the evolution of VO and OV alternation in main and subordinate clauses with a monotransitive verb and an overt subject; (b) examine whether the development of the VO and OV alternation in Romeyka is (i) the result of internal (endogenous) change or continuity, or rather (ii) the result of external change due to contact with Turkish; and, (c) I propose a feature-based theory which makes predictions on language change due to contact within the generative framework (see Chomsky 1995 et seq.).

Romeyka[1] is the only remaining variety of AMG that is still spoken in the area historically known as Asia Minor (present-day Anatolia, Turkey). In fact, today there remain three Greek-speaking enclaves: Of/Çaykara, Sürmene and Tonya (see Neocleous 2020: 11 with further references).

---

[1] For a discussion on the name 'Romeyka', see Schreiber and Sitaridou (2017: 2).

Nicolaos Neocleous: Queens' College, Silver Street, Cambridge CB3 9ET, United Kingdom, E-mail: nicneocleous@hotmail.com

The article is structured as follows: in Section 2, I discuss the methodology I use in this study; Section 3 explores the evolution of VO and OV alternation in main clauses in Romeyka; and Section 4 depicts the evolution of VO and OV alternation in subordinate clauses in Romeyka. The main findings of the article are presented in Section 5.

2 Methodology

The results reported here were obtained from:

(a) A corpus consisting of Romeyka and Turkish data collected during fieldwork I carried out in July 2015 in a remote part of the Of/Çaykara region, in a village, which will be referred to as 'Anasta', in order to preserve the anonymity of the informants and the village (Sitaridou 2014a: 29, fn. 3). The corpus comprises 06:48:21 hours of audio recordings (see Neocleous 2020). My main informants were two women aged 48 and 74 at the time of the fieldtrip, two men aged 37 and 46 and two boys aged 12 and 14. The data collection entailed oral interviews based on structured questionnaires, as well as spontaneous data. Because the direct approach to fieldwork with questionnaire-based data elicitation could have had an effect on the data especially in subordinate clauses which are in general difficult to elicit; for example, the language of elicitation could have had an effect. In order to avoid those biases, I conducted the fieldwork in Romeyka to elicit Romeyka data and in Turkish to elicit Turkish data. A pilot test of the questionnaires was first carried out with a Turkish native speaker (speaking the variety of Istanbul, which is said to represent the Standard Modern Turkish variety) and a Greek native speaker (speaking the variety of Athens, which is said to represent the Standard Modern Greek variety). The data were audio recorded, transcribed and annotated.

(b) Published works that include diachronic data other than Romeyka or Turkish spoken in 'Anasta'. The textual sources are cited in each linguistic example and are presented in detail in a separate section in the end of this article (see Section Textual sources).

In this study, I essentially follow Sitaridou's (2016) reconstruction method. I compare the VO and OV alternation in main and subordinate clauses in Romeyka with the VO and OV alternation in (a) HelGr, (b) MedGr and (c) Anasta Turkish. Depending on the findings of this comparison, I make two predictions: First, if VO and OV alternation in Romeyka is similar to that found in previous stages of Greek and not to that found in Anasta Turkish, then we can safely conclude that this is the result of continuity from previous stages of Greek. Seccond, if the VO and OV alternation in Romeyka is similar to that found in Anasta Turkish and not to that found in previous stages of Greek, then we can conclude that this is the result of external change due to contact with Anasta Turkish.

Since word order variation in Romeyka is argued to be discourse-driven (see Neocleous 2020: chapter 4), discourse features are taken into account as follows:

(a) A pragmatically unmarked order is said to be an 'all-focus sentence', *aka* 'a presentational focus sentence', containing neither old information nor any presuppositions. Several diagnostics can be applied to determine the canonical order in a language. The first one is a 'what happened?' question, which typically invokes a context in which all of the elements of the answer constitute new information and hence are equal in terms of their discourse-pragmatic properties (Büring 2009; van der Wal 2016).

(b) A pragmatically marked order is said to be a sentence in which at least one constituent is 'focused'. Throughout this study, I adopt the semantic definition of focus proposed by Rooth's Alternative Semantics (1985, 1992, 1996), which states that focus indicates the presence of alternatives that are relevant for the interpretation of linguistic expressions. The information focus asserts the membership of an individual in a set (see Gundel 1998). The most widespread and accepted test for focus and a method of establishing the scope of focus is *wh*-questions and their answers (Beaver and Clark 2008; Krifka 2007; Lambrecht 1994; Rooth 1992; van der Wal 2016, i.a.). The basic idea is that a *wh*-question always yields new information. If focus is defined as the new information in a sentence, then it follows that the phrase that replaces the *wh*-constituent is focused. Contrastive focus involves the selection of a subset from a set of alternatives (see Molnár 2006).

3 The evolution of VO and OV alternation in main clauses in Romeyka

I begin my investigation with VO and OV alternation in main clauses in Romeyka. To start with, the pragmatically unmarked word order in main clauses in Romeyka is VO (see 1), while OV occurs when the object is focused, regardless of whether it is information focus (see 2) or contrastive focus (see 3):

(1) Romeyka [personal corpus]
    a. Question:
       do   *ejéndo?*
       what.NOM   happen.PST.3SG
       'What happened?'

b. Answer:
   o   mustafás   epelǽpsen   to   χoráfin.
   the.NOM   Mustafas.NOM put.fertiliser.PST.3SG the.ACC field.ACC
   'Mustafas manured the field.'

(2)   Romeyka   [personal corpus]
   a. Question:
      alís   dóyna   éfaen?
      Alis.NOM   what.ACC   eat.PST.3SG
      'What did Alis eat?'
   b. Answer:
      alís   [χavíts]_{I-Foc}   éfaen.
      Alis.NOM   pudding.ACC   eat.PST.3SG
      'Alis ate a pudding.'

(3)   Romeyka   [personal corpus]
   a. Question:
      kahvén   jóksa   tšáin   θélis?
      coffee.ACC or   tea.ACC   want.2SG
      'Do you want coffee or tea?'
   b. Answer:
      eγó   [kahvén]_{C-Foc}   θélo.
      I.NOM   coffee.ACC   want.1SG
      'I want coffee.'

Likewise, in HelGr, the pragmatically unmarked word order in main clauses is VO (Kirk 2012: 41) (see 4), while OV results from movement of the focused object, regardless of whether it is information focus (see 5) or contrastive focus (see 6):

(4)   Hellenistic Greek (transliteration and glosses were adapted)
      ánthropós   tis   epoíe:se   deîpnon   méga.
      man.NOM   INDEF.NOM   make.PST.3SG   dinner.ACC   large.ACC
      'A certain man made a large dinner.'   [NT, Lk 14:16][2]

(5)   Hellenistic Greek (transliteration and glosses were adapted)
      ei   emè   é:ideite   kaì   [tòn   patéra   mou]_{I-Foc}
      if   I.ACC   know.PNP.2PL   and   the.ACC   father.ACC   I.POSS

---

[2] NT = New Testament. The citation of the examples from the NT follows the "Common Abbreviations for Books of the *Bible –Turabian* (The Chicago Manual of Style)", i.e., Jn = *Gospel of John*, Lk = *Gospel of Luke*, Mk = *Gospel of Mark*, Mt = *Gospel of Matthew*, followed by the chapter of the book and the paragraph, e.g., Lk 14:16 = *Gospel of Luke*, chapter 14, paragraph 16.

  án   é:ideite.
  PCL   know.PNP.2PL
  'If you had known me, you would also have known my father.'
                     [NT, Jn 8:19]

(6) Hellenistic Greek (transliteration and glosses were adapted)
  *[éleos]*_{C-Foc}   *thélo: kaì   ou   thusían.*
  mercy.ACC   want.1SG   and   NEG   sacrifice.ACC
  'I want mercy and not sacrifice.'   [NT, Mt 9:13, 12:7]

Unlike in Romeyka and HelGr, in MedGr, VO is both pragmatically unmarked and marked. For instance, for MedGr see the pragmatically unmarked VO order in (7), an *in-situ* information focused object in (8), as well as an *in-situ* contrastive focused object in (9):

(7) Medieval Greek (transliteration and glosses were adapted)
  patéra   mu,   i   θiúðes   tu   riγós   ívran   ton
  father   I.POSS   the   uncles.NOM   the   king.GEN   find.PST.3SG   he.ACC
  orfanón   ce   ptoχón,   ce   emirástisan   to
  orphan.ACC   and   poor.ACC   and   divide.PST.3PL   the.ACC
  nis:ín   tu.
  island.ACC   he.POSS
  'Oh my father, the uncles of the king found him orphaned and poor and they divided his island.'   [Chronicle of Machairas: 468]

(8) Medieval Greek (transliteration and glosses were adapted)
  *íksevre   [ce   túton]*_{I-Foc}*.*
  know.IMPF.3SG   PCL   this.ACC
  'He also knew that.'   [Chronicle of Machairas: 6]

(9) Medieval Greek (transliteration and glosses were adapted)
  ce   pépse   frénimus   mandatofórus;
  and   send.IMP.2SG   restrained.ACC   messengers.ACC
  *miðén   pépsis   [atíχus]*_{C-Foc}*.*
  NEG   send.PNP.2SG   impetuous.ACC
  'Also, send restrained messengers; do not send impetuous (messengers).'
               [Chronicle of Machairas: 25]

In contrast to Romeyka, in Anasta Turkish, the pragmatically unmarked word order is OV (see 10):

(10) Anasta Turkish [personal corpus]
a. Question:
Dün   ne   yap-tı-nız?
yesterday what   do-PST-2PL
'What did you do yesterday?'
b. Answer:
Rumca   konuş-tu-k.   Yaz-dı-k.   Otur-du-k.
Romeyka   talk-PST-1PL   write-PST-1PL   sit-PST-1PL
Televizyone   izle-di-k.   yemek   ye-di-k.
television   watch-PST-1PL   food   eat-PST-1PL
Çay   iş-ti-k.
tea   drink-PST-1PL
'We talked and wrote in Romeyka. We sat down and watched television. We had a meal and tea.'

Unlike in Romeyka, in Anasta Turkish OV also occurs when the object is focused (see 11):

(11) Anasta Turkish [personal corpus]
a. Question:
Ali   bir   kere   kim-i   öp-tü?
Ali   one   time   who-ACC   kiss-PST.3SG
'Who did Ali kiss once?'
b. Answer:
Ali   bir   kere   [Ayşe-'yi]$_{\text{I-Foc}}$ öp-tü.
Ali   one   time   Ayşe-ACC   kiss-PST.3SG
'Ali kissed Ayşe once.'

Additional data can be drawn from the Anatolian Turkish variety of Cappadocia (Karamanlidika),[3] which is genetically related to the one spoken in 'Anasta' and bears the same structure to Standard Modern Turkish, rather than that employed by HelGr. The pragmatically unmarked order in Karamanlidika Turkish

---

[3] Karamanlidika is a Turkish variety that was spoken by (a) monolingual Greek Orthodox communities in Cappadocia and (b) bilingual (in Karamanlidika and Cappadocian Greek) Greek Orthodox communities in Cappadocia before their relocation to Greece due to the Treaty of Lausanne for the Population Exchange between Greece and Turkey (1923). It is said that both Cappadocian and Pontic Greek are phylogenetically related and belong to the Asia Minor Greek group of Greek varieties (see Sitaridou 2016). Since both Anasta Turkish and Karamanlidika Turkish are spoken by bilingual communities (in Anatolian Turkish and Asia Minor Greek), it is predicted that both Turkish varieties share similar grammatical features. If that is the case, a comparison between them could shed light to this study.

is OV (see 12a), like in Standard Modern Turkish (see 12b) and unlike in HelGr, in which it is VO (see 4):

(12) Pragmatically unmarked order
    a. Karamanlidika Turkish (transliteration and glosses were adapted)
        Bir   adam   azim   meclis   eyle-di.
        a   man   large   gathering   give-PST.3SG
        'A certain man was preparing a great banquet.'   [NT, Lk 14:16]
    b. Standard Modern Turkish (transliteration and glosses were adapted)
        Adam-ın   bir-i   büyük   bir   şölen   hazırla-yıp
        man-GEN   one-POSS-3SG   large   one   gathering   prepare-CVB
        'A certain man was preparing a great banquet and....'   [NT, Lk 14:16]

As for the information focus, Karamanlidika Turkish manifests OV (see 13a), like in Standard Modern Turkish (see 13b) and HelGr (see 5):

(13) Information focus
    a. Karamanlidika Turkish (transliteration and glosses were adapted)
        eğer   ben-i   bil-e-idi-niz
        if   I-ACC   know-OPT-PST-2PL
        [peder-im-i]$_{I\text{-}Foc}$ dahı   bil-ir=idi-niz.
        father-I.POSS-ACC   also   know-AOR=PST-2PL
        'If you knew me, you would know my Father also.'   [NT, Jn 8:19]
    b. Standard Modern Turkish (transliteration and glosses were adapted)
        Ben-i   tanış-a-ydı-nız,
        I-ACC   know-OPT-PST-2PL
        [Baba-m-ı]$_{I\text{-}Foc}$ da   tanı-r=dı-nız.
        father-I.POSS-ACC   also   know-AOR=PST-2PL
        'If you knew me, you would know my Father also.'   [NT, Jn 8:19]

Regarding the contrastive focus, Karamanlidika Turkish bears OV (see 14a), like in Standard Modern Turkish (see 14b) and HelGr (see 6):

(14) Contrastive focus
    a. Karamanlidika Turkish (transliteration and glosses were adapted)
        [Merhamet]$_{C\text{-}Foc}$ iste-r-im,   ve   kurban   iste-me-m.
        mercy   want-AOR-1SG   and   sacrifice   want-NEG-1SG
        'I desire mercy, not sacrifice.'   [NT, Mt 9:13, 12:7]
    b. Standard Modern Turkish (transliteration and glosses were adapted)
        Ben   kurban   değil,   [merhamet]$_{C\text{-}Foc}$ iste-r-im.
        I   sacrifice   NEG   mercy   want-AOR-1SG
        'I desire mercy, not sacrifice.'   [NT, Mt 9:13, 12:7]

Overall, our data show that the pragmatically unmarked word order in Romeyka is VO as with HelGr and MedGr and different from Anasta Turkish. This suggests that the pragmatically unmarked word order in Romeyka derives from previous stages of Greek and not from Turkish. However, due to the same form of information focus and contrastive focus in Romeyka, HelGr and Anasta Turkish, it is difficult to come up with a safe conclusion regarding their diachronic development.

Additional evidence derives from the fact that the word order in a variety of Laz, a Kartvelian language spoken in an area close to 'Anasta', i.e., in Pazar, is not the same as that in Georgian, a language to which it is genetically related (see Öztürk and Pöchtrager 2011). On the contrary, it is similar to Turkish. Note that Pazar Laz is in close contact with Turkish.

In Laz, the pragmatically unmarked word order is OV, like in Anasta Turkish (see 15):

(15)   Pazar Laz (transliteration and glosses were adapted)
    a. Question:
        'What happened?'
    b. Answer:
        Alik   çitabi   dót'k'u.
        Ali.NOM   book.ACC   read.PST.3SG
        'Ali read the book.'   [Göksel 2011: 146]

The information focus in Laz results in OV orders, like in Anasta Turkish (see 16):

(16)   Pazar Laz (transliteration and glosses were adapted)
        Alik   çitabi   [Ayşes]$_{I-Foc}$ kómeçu.
        Ali.NOM   book.ACC   Ayşe.DAT   give.PST.3SG
        'Ali gave the book to Ayşe.'   [Göksel 2011: 149]

Finally, the contrastive focus in Pazar Laz results in OV orders, like in Anasta Turkish (see 17):

(17)   Pazar Laz (transliteration and glosses were adapted)
        Alik   çitabi   [Ayşes]$_{C-Foc}$ kómeçu.
        Ali.NOM   book.ACC   Ayşe.DAT   give.PST.3SG
        'Ali gave the book to Ayşe.'   [Göksel 2011: 149]

If the pragmatically unmarked word order in main clauses in Romeyka had changed due to its contact with Anasta Turkish, we would have expected OV to be the pragmatically unmarked word order in Romeyka like in Laz. Subsequently, the pragmatically unmarked word order in Pazar Laz is a clue in favor of a non-contact explanation for the Romeyka counterpart, since Pazar Laz must have reshaped its word order schema in accordance with the Turkish one. This is a strong indication that the pragmatically unmarked VO order in Romeyka derives from previous stages of Greek.

Under those circumstances, it becomes obvious that the pragmatically unmarked order in main clauses in Romeyka developed from previous stages of Greek (see Table 1):

Table 1: Inherited development (Sitaridou 2016).

a.	In Hellenistic Greek, the pragmatically unmarked word order is VO with OV resulting from movement of the focused object
b.	In Medieval Greek, the pragmatically unmarked word order is VO with focus *in-situ*
c.	In Anasta Turkish, the pragmatically unmarked word order is OV with the focused object *in-situ*
d.	In Romeyka, the pragmatically unmarked word order is VO with OV resulting from movement of the focused object
e.	The pragmatically unmarked VO order in Romeyka descends either from Hellenistic Greek or Medieval Greek
f.	Therefore, the pragmatically unmarked VO order is inherited (from previous stages of Greek)

Crucially, data from other AMG varieties, namely Cappadocian, Phárasiot, Pontic and Sílliot, support this analysis. In all AMG varieties, the pragmatically unmarked word order is VO, like in Romeyka. However, except for Pontic Greek, the organization of information structure in Cappadocian, Phárasiot and Sílliot more closely resembles the one found in MedGr, rather than the one in Romeyka (see Table 2):

Table 2: Distribution of information structure in AMG varieties.

	Romeyka	Pontic Greek	Cappadocian	Phárasiot	Sílliot
Unmarked	SVO	SVO	SVO	SVO	SVO
I-Foc	S[O]V	S[O]V	SV[O]	SV[O]	SV[O] S[O]V
C-Foc	S[O]V	S[O]V	n/a	SV[O]	n/a

In a nutshell, the distribution of VO and OV alternation in other AMG varieties indicates that: (a) the pragmatically unmarked word order in AMG varieties rives from previous stages of Greek, (b) the information focus and contrastive focus in Cappadocian, Phárasiot and Sílliot could derive from MedGr, and (c) it is still unclear whether the information focus and contrastive focus in Romeyka and Pontic Greek derive (i) from HelGr, or (ii) from MedGr, and (iii) whether it had reshaped after the Turkish equivalent due to its contact with it.

4 The evolution of VO and OV alternation in subordinate clauses in Romeyka

In Romeyka, the pragmatically unmarked word order in subordinate clauses is OV (see 18). OV is attested too when the object is focused, regardless of whether it is information focus (see 19) or contrastive focus (see 20):

(18) Romeyka [personal corpus]
 eγó θaró, alís c^{h}it^{h}ápæ eχújepsen.
 I.NOM think.1SG Alis.NOM books.ACC read.PST.3SG
 'I think that Alis read books.'

(19) Romeyka [personal corpus]
 a. Question:
 do θarís, alís tínan efílisen?
 what.ACC think.2SG Alis.NOM who.ACC kiss.PST.3SG
 'Who do you think that Alis kissed?'
 b. Answer:
 eγó θaró, alís [tin aišén]$_{I\text{-}Foc}$ efílisen.
 I.NOM think.1SG Alis.NOM the.ACC Ayşe.ACC kiss.PST.3SG
 'I think that Alis kissed Ayşe.'

(20) Romeyka [personal corpus]
 a. Question:
 alís dóyna éfaen?
 Alis.NOM what.ACC eat.PST.3SG
 mílon éfaen i ap^{h}íðin éfaen?
 apple.ACC eat.PST.3SG or pear.ACC eat.PST.3SG
 'What did Alis eat? An apple or a pear?'
 b. Answer:
 eγó léγo, alís [ap^{h}íðin]$_{C\text{-}Foc}$ éfaen.
 I.NOM say.1SG Alis.NOM pear.ACC eat.PST.3SG
 'I say that it's a pear that Alis ate.'

Unlike in Romeyka, in HelGr the pragmatically unmarked word order in subordinate clauses is VO (see 21) and the same applies to MedGr (see 22):

(21) Hellenistic Greek (transliteration and glosses were adapted)
... légo:n   hóti   entrapé:sontai   tòn   hyión   mou.
say.PTCP   that   respect.3PL   the.ACC   son.ACC   I.POSS
'... saying that they will respect my son.'   [Mk 12:6]

(22) Medieval Greek (transliteration and glosses were adapted)
... lalónta   tus   óti   éxomen   kalín   ayápin   meson   mas;
say.GER   them   that   have.1PL   good.ACC   love.ACC   between   us
'... saying to them that we have peace between us.'
[Chronicle of Machairas: 230]

In Anasta Turkish, the pragmatically unmarked word order in subordinate clauses is OV, like in Romeyka (see 23):

(23) Anasta Turkish   [personal corpus]
Hiç   iste-mi-yor-um   yemek   yap-ma-yacağ-ım.
no   want-NEG-PROG-1SG   food   make-NEG-FUT-1SG
'I don't want to make food.'

Since the pragmatically unmarked word order in subordinate clauses in Romeyka is OV, like that in Anasta Turkish and unlike that in HelGr and MedGr, I argue that OV in subordinate clauses in Romeyka is the result of its contact with Turkish.

In order to account for the grammatical mechanism that have triggered such a change in the pragmatically unmarked word order in subordinate clauses in Romeyka, I apply the feature-based analysis, which was developed in Neocleous (2020) and Neocleous and Sitaridou (2022). In particular, I build my proposal on Tsimpli and Dimitrakopoulou (2007) and Tsimpli and Mastropavlou's (2007) distinction between formal features that are visible at the syntax-semantics interface because of their semantic import, i.e., SEM-interpretable features, and those whose role is restricted to syntactic derivations and possibly have PHON-realization but no role at SEM, i.e., the SEM-uninterpretable features. Based on this distinction, there are the following (un)interpretability possibilities between SEM and PHON (see 24):

(24)   (Un)interpretability possibilities between SEM and PHON:
   a.   SEM-interpretable/PHON-uninterpretable features (e.g., animacy distinctions on Greek nouns and pronouns are not grammaticalised due to grammatical gender differences)

b. SEM-interpretable/PHON-interpretable (e.g., animacy distinctions on English *wh-* and personal pronouns)
c. SEM-uninterpretable/PHON-interpretable (e.g., resumptive uses of subject–verb agreement and object clitics in Greek)
d. SEM-uninterpretable/PHON-uninterpretable (e.g., case and subject–verb agreement in English) [Tsimpli and Dimitrakopoulou 2007: 223]

Neocleous (2020: chapter 4) has shown that Romeyka word order variation is discourse-driven. Further, Neocleous (2020) argues that discourse-related features are encoded in a formal feature, i.e., the linearization feature. The linearization feature drives the computation in the narrow syntax and maps syntactic units (phases) as logical forms (LFs) onto the SEM and as phonological forms (PFs) onto the PHON. The pragmatic interpretation of the clause takes place at the SEM, in which the linear order of the constituents plays a vital role in the interpretation of their pragmatic value.

That is to say, for the linearization feature there must be the following (un)interpretability possibilities between the SEM and the PHON (see 25):

(25) (Un)interpretability possibilities between the SEM and the PHON of the linearization feature:
  a. SEM-interpretable/PHON-uninterpretable (VO, the object is focused) (e.g., see 26)
  b. SEM-interpretable/PHON-interpretable (OV, the object is focused) (e.g., see 27)
  c. SEM-uninterpretable/PHON-interpretable (OV, the object is not focused) (e.g., see 28)
  d. SEM-uninterpretable/PHON-uninterpretable (VO, the object is not focused) (e.g., see 29)

(26) Modern Greek (transliteration and glosses were adapted)
  a. Question:
  ti   éfaje   o   jóryos?
  what.ACC eat.PST.3SG the.NOM   George.NOM
  'What did George eat?'
  b. Answer:
  o   jóryos   éfaje   [tin   kobósta]$_{\text{I-Foc}}$.
  the.NOM   George.NOM   eat.PST.3SG   the.ACC   stewed-fruit.ACC
  'George ate the stewed fruits.'   [Sitaridou and Kaltsa 2014: 12]

(27) Turkish (transliteration and glosses were adapted)
  a. Question:
  Filiz-in   kardeş-ler-i   ne   iç-t-i   parti–de?
  Filiz-GEN   sister-PL-POSS   what   drink-PST-3SG   party-LOC
  'What did Filiz's sisters get to drink at the party?'

b. Answer:
Valla tüm kardeş-ler-den haber-i-m yok, ama
frankly all sister-PL-ABL news-POSS-1SG NEG but
Filiz-in en küçük kardeş-i [rakı-dan]$_{I\text{-Foc}}$ iç-t-i.
Filiz-GEN most young sister-POSS rakı-ABL drink-PST-3SG
'Frankly, I do not know about all the sisters but Filiz's youngest sister drank (from the) rakı.' [Şener 2010: 35]

(28) Turkish (transliteration and glosses were adapted)
a. Question:
Ne ol-d-u?
what happen-PST-3SG
'What happened?'
b. Answer:
Cadı hırsız-ı lanetle-d-i.
witch thief-ACC curse-PST-3SG
'The witch cursed the thief.' [Şener 2010: 10]

(29) Modern Greek (transliteration and glosses were adapted)
a. Question:
'What happened?'
b. Answer:
éspase ti lába o jánis.
break.PST.3SG the.ACC lamp.ACC the.NOM Yanis.NOM
'Yanis broke the lamp.' [Alexiadou and Anagnstopoulou 2000: 174]

In other words, what differentiates a pragmatically unmarked VO order (see 25a) from a pragmatically unmarked OV order (see 25c) is (a) the presence of the linearization feature in the latter, but not in the former, and (b) the lack of internal semantic interpretation in the linearization feature in the latter (i.e., SEM-uninterpretable).

As for the distribution of the linearization feature in subordinate clauses in Romeyka, I argue that the linearization feature in focused OV orders is SEM-interpretable and PHON-interpretable (see 30), whereas in nonfocused OV orders it is SEM-uninterpretable and PHON-interpretable (see 31) (see Table 3).

(30) Romeyka [personal corpus]
eyó θaró, alís [tin aišén]$_{I\text{-Foc}}$ efílisen.
I.NOM think.1SG Alis.NOM the.ACC Ayşe.ACC kiss.PST.3SG
'I think that Alis kissed Ayşe.'

(31) Romeyka [personal corpus]
 eyó θaró, alís tin aišén efílisen.
 I.NOM think.1SG Alis.NOM the.ACC Ayşe.ACC kiss.PST.3SG
 'I think that Alis kissed Ayşe.'

Table 3: Interface (un)interpretability of the linearization feature in focused and nonfocused order subordinate clauses in Romeyka.

| Focused | SEM-interpretable | PHON-interpretable |
| Nonfocused | SEM-uninterpretable | PHON-interpretable |

In Anasta Turkish, I argue that like in Romeyka, the linearization feature in focused OV orders is SEM-interpretable and PHON-interpretable (see 32), whereas in nonfocused OV orders it is SEM-uninterpretable and PHON-interpretable (see 33) (see Table 4).

(32) Anasta Turkish [personal corpus]
 Hiç iste-mi-yor-um [yemek]$_{I\text{-}Foc}$ yap-ma-yacağ-ım.
 no want-NEG-PROG-1SG food make-NEG-FUT-1SG
 'I don't want to make food.'

(33) Anasta Turkish [personal corpus]
 Hiç iste-mi-yor-um yemek yap-ma-yacağ-ım.
 no want-NEG-PROG-1SG food make-NEG-FUT-1SG
 'I don't want to make food.'

Table 4: Interface (un)interpretability of the linearization feature in focused and nonfocused order subordinate clauses in Anasta Turkish.

| Focused | SEM-interpretable | PHON-interpretable |
| Nonfocused | SEM-uninterpretable | PHON-interpretable |

Essentially the problem is, what is the initial state grammar that yielded the current state in Romeyka? In particular, in an early stage, I argue that the linearization feature in focused OV orders was SEM-interpretable and PHON-interpretable (see 34), whereas in nonfocused VO orders it was SEM-uninterpretable and PHON-uninterpretable (see 35) (see Table 5).

(34) Romeyka
*eyó θaró,   alís   [tin   aišén]_I-Foc   efílisen.
I.NOM think.1SG   Alis.NOM   the.ACC   Ayşe.ACC   kiss.PST.3SG
'I think that Alis kissed Ayşe.'

(35) Romeyka
*eyó   θaró,   alís   efílisen   tin   aišén.
I.NOM   think.1SG   Alis.NOM   kiss.PST.3SG   the.ACC   Ayşe.ACC
'I think that Alis kissed Ayşe.'

Table 5: Early stage of the interface (un)interpretability of the linearization feature in focused and nonfocused order subordinate clauses in Romeyka.

Focused	SEM-interpretable	PHON-interpretable
Nonfocused	SEM-uninterpretable	PHON-uninterpretable

I therefore assume that the pragmatically unmarked VO order in subordinate clauses in Romeyka shifted to OV as a result of language contact with Turkish. That is to say, the trajectory word order change in subordinate clauses in Romeyka is shown in (36):

(36) SEM-uninterpretable/PHON-uninterpretable (VO, the object is not focused) (see (37))
develops to
SEM-uninterpretable/PHON-interpretable (OV, the object is not focused) (see (38))

(37) Romeyka
*eyó   θaró,   alís   efílisen   tin   aišén.
I.NOM   think.1SG   Alis.NOM   kiss.PST.3SG   the.ACC   Ayşe.ACC
'I think that Alis kissed Ayşe.'

(38) Romeyka   [personal corpus]
eyó   θaró,   alís   tin   aišén   efílisen.
I.NOM   think.1SG   Alis.NOM   the.ACC   Ayşe.ACC   kiss.PST.3SG
'I think that Alis kissed Ayşe.'

According to (36), the original absence of the linearization feature in Romeyka nonfocused orders (SEM- and PHON-uninterpretable) changed into the appearance of a linearization feature without semantic structure (SEM-uninterpretable) because of its contact with the equivalent Anasta Turkish structure.

In order to resolve the contact trajectory, I depart from the following premises on contact-induced syntactic change:

First, copying was possible because Romeyka allowed both orders to start with. That is, interference due to language contact was triggered in Romeyka, because both parameter values of the linearization feature existed in both Romeyka and Anasta Turkish. This supports the Resistance Principle (Guardiano et al. 2016: 54) (see 39):

(39) Resistance Principle:
Resetting of parameter α from value X to Y in language A as triggered by interference of language B only takes place if a subset of the strings that contribute to constituting a trigger for value Y of parameter α in language B already exists in language A. [Guardiano et al. 2016: 54]

In other words, the resetting of a parameter under the influence of interference data is possible only if the new triggers are similar enough to triggers already unmistakably present in the interfered language, though of course not sufficient on their own to trigger the new value (Guardiano et al. 2016: 54). The informal idea is that interference data in parametric syntax must appear at least in part as "familiar" in the interfered language, in order to be used as triggers; thus "contact may exacerbate/reinforce existing tendencies" (Sitaridou 2014a).

Second, I pursue a theoretical approach in that in multilingual environments it is the SEM-uninterpretable features of a language A that are not instantiated in the language B or *vice versa* that cause learnability problems. I therefore assume transfer from Turkish into Romeyka whereby the SEM-uninterpretable features of Romeyka cause learnability problems. Based on these assumptions, I propose a generalization of contact-induced word order change in Romeyka (see 40):

(40) Generalization of contact-induced change in word order in Romeyka:
The PHON-realization of the linearization feature which is SEM-uninterpretable is sensitive to contact-induced change.

Third and equally important, as the Input Generalization states, acquirers tend to generalize the input from the above alternations; children conclude that the linearization feature must always be present, even if there is no semantic interpretation (see 41):

(41) Input Generalization:
If a functional headsets parameter p_j to value v_i then there is a preference for similar functional heads to set p_j to value v_i.
[Biberauer and Roberts 2015]

This account explains the directionality of cross-linguistic effects: it is always the language that instantiates the more restrictive option that affects the other, not *vice versa* (see Feature Economy in Biberauer and Roberts 2015). Hence, it is Anasta Turkish that affects Romeyka regardless of whether the latter is the L1 (attrited or heritage) or L2.

In the final analysis, as shown above, the pragmatically unmarked OV order in subordinate clauses in Romeyka is assumed to have been reshaped after Anasta Turkish (see Table 6):

Table 6: Contact-induced change (Sitaridou 2016).

a.	In Hellenistic Greek, the pragmatically unmarked word order is VO
b.	In Medieval Greek, the pragmatically unmarked word order is VO
c.	In Anasta Turkish, the pragmatically unmarked word order is OV
d.	In Romeyka, the pragmatically unmarked word order is OV
e.	Therefore, Romeyka OV is contact induced by Anasta Turkish

5 Conclusions

In this article, I investigated the evolution of VO and OV alternation in main and subordinate clauses in Romeyka. The main findings of the article indicate that (a) the pragmatically unmarked VO order in main clauses in Romeyka is the result of continuity from previous stages of Greek and (b) the pragmatically unmarked OV order in subordinate clauses in Romeyka is the result of external change due to contact with Anasta Turkish. Finally, I provided a principled explanation of word order change due to language contact, arguing that the SEM-uninterpretable features are sensitive to change their PHON-realization due to language contact. That said, SEM-uninterpretable features, such as agreement features and anaphoric elements are predicted to be sensitive to change at PHON due to language contact. In conclusion, if the arguments in the present study are on the right track, the predictions that have been made will need to be tested cross-linguistically, both synchronically and diachronically, in order to evaluate their validity. Obviously, only future work can tell us whether the views in the present study can be sustained or not.

Acknowledgments: I am grateful to the audiences of the conference on *Language Contact and Language Change in Western Asia (LCaLCiWA)* (Goethe University Frankfurt, Germany, 10-12 March 2017) and the Spring School *Contact Linguistics in Cross-border Kurdistan (CLiCK)* (13–17 March 2017, Goethe University Frankfurt, Germany); their suggestions and comments led to considerable improvements. I also wish to thank the A. G. Leventis Foundation and the Aridemi Stiftung for their financial support of my research, as well as the Lister Fund granted to me by Queens' College, University of Cambridge and funding for work carried out of Cambridge granted to me by the Faculty of Modern and Medieval Languages and Linguistics, University of Cambridge, thanks to which I collected the data presented in this study during fieldwork I conducted in Pontus. To Dr Ioanna Sitaridou should be ascribed in large part such merits as this study may claim. Needless to say, all the errors are my own.

Abbreviations

Published examples

Jn	Gospel of John
Lk	Gospel of Luke
Mk	Gospel of Mark
Mt	Gospel of Matthew
NT	New Testament

Languages

AMG	Asia Minor Greek
HelGr	Hellenistic Greek
MedGr	Medieval Greek

Glosses

1, 2, 3	1st, 2nd, 3rd person
ABL	ablative case
ACC	accusative case
AOR	aorist
C-FOC	contrastive focus
CVB	converb
DAT	dative case
FUT	future tense

GEN	genitive case
GER	gerund
IMPF	imperfect
INDEF	indefinite
I-FOC	information focus
L1, 2	1st, 2nd language
LOC	locative case
NEG	negation marker
NOM	nominative case
O	object
OPT	optative mood
PTCP	participle
PST	past tense
PCL	particle
PHON	phonological component
PL	plural number
PNP	perfective aspect, nonpast tense
POSS	possessive
PROG	progressive
S	subject
SEM	semantic component
SG	singular number
V	verb

References

Alexiadou, Artemis & Elena Anagnstopoulou. 2000. Greek syntax: A principles and parameters perspective. *Journal of Greek Linguistics* 1. 169–221.

Beaver, David & Brady Clark. 2008. *Sense and sensitivity*. Oxford: Blackwell.

Biberauer, Theresa & Ian Roberts. 2015. Rethinking formal hierarchies: A proposed unification. *Cambridge Occasional Papers in Linguistics* 7. 1–31.

Biberauer, Theresa, Anders Holmber & Ian Roberts. 2014. A syntactic universal and its consequences. *Linguistic Inquiry* 45(2). 169–225.

Büring, Daniel. 2009. Towards a typology of focus realization. In Malte Zimmermann & Caroline Féry (eds.), *Information structure*, 177–205. Oxford: Ofxord University Press.

Chomsky, Noam. 1995. *The Minimalist Program*. Cambridge, MA: MIT Press.

Göksel, Aslı. 2011. Word order, sentential stress and intonation. In Balkız Öztürk & Markus A. Pöchtrager (eds.), *Pazar Laz*, 146–159. Munich: Lincom.

Guardiano, Cristina, Dimitris Michelioudakis, Andrea Ceolin, Monica-Alexandrina Irimia, Giuseppe Longobardi, Nina Radkevich, Giuseppina Silvestri & Ioanna Sitaridou. 2016. South by Southeast. A syntactic approach to Greek and Romance microvariation. *L'Italia dialettale* LXXVII. 95–166.

Gundel, Janette K. 1998. On three kinds of focus. In Peter Bosch & Rob van der Sandt (eds.), *Focus: Linguistic, cognitive and computational perspectives*, 293–305. Cambridge: Cambridge University Press.
Kirk, Alison. 2012. *Word order and information structure in New Testament Greek*. Leiden: University of Leiden Ph.D. Thesis.
Krifka, Manfred. 2007. The semantics of questions and the focusation of answers. In Chingmin Lee, Matthew Gordon & Daniel Büring (eds.), *Topic and focus, crosslinguistic perspectives on meaning and intonation*, 139–150. Dordrecht: Springer.
Lambrecht, Knud. 1994. *Information structure and sentence form*. Cambridge: Cambridge University Press.
Molnár, Valéria. 2006. On different kinds of contrast. In Valéria Molnár & Susanne Winkler (eds.), *Architecture of focus. Studies in generative grammar*, vol. 82, 197–233. Berlin & New York: De Gruyter.
Neocleous, Nicolaos. 2020. *Word order and information structure in Romeyka: A syntax and semantics interface account of order in a minimalist system*. Cambridge: University of Cambridge Ph.D. Thesis.
Neocleous, Nicolaos & Ioanna Sitaridou. 2022. Never just contact. The rise of final auxiliaries in Asia Minor Greek. *Diachronica*. [DOI:https://doi.org/10.1075/dia.17048}].
Öztürk, Balkız & Markus A. Pöchtrager. 2011. *Pazar Laz*. Muenchen: Lincom Europa.
Rooth, Mats. 1985. *Association with focus*. Amherst, MA, USA: University of Massachussets doctoral dissertation.
Rooth, Mats. 1992. A theory of focus interpretation. *Natural Language Semantics* 1(1). 75–116.
Rooth, Mats. 1996. Focus. In Shalom Lappin (ed.), *The handbook of contemporary semantic theory*, 271–298. Oxford: Blackwell.
Schreiber, Laurentia & Ioanna Sitaridou. 2017. Assessing the sociolinguistic vitality of Istanbulite Romeyka: An attitudinal study. *Journal of Multilingual and Multicultural Development* 39(1). 1–16.
Şener, Serkan. 2010. *(Non-)peripheral matters in Turkish syntax*. Connecticut: University of Connecticut Ph.D. Thesis.
Sitaridou, Ioanna & Maria Kaltsa. 2014. Contrastivity in Pontic Greek. *Lingua* 146. 1–27.
Sitaridou, Ioanna. 2013. Greek-speaking enclaves in Pontus today: The documentation and revitalization of Romeyka. In Mari C. Jones & Sarah Ogilvie (eds.), *Keeping languages alive. Language endangerment: Documentation, pedagogy and revitalization*, 98–112. Cambridge: Cambridge University Press.
Sitaridou, Ioanna. 2014a. The Romeyka infinitive: Continuity, contact and change in the Hellenic varieties of Pontus. *Diachronica* 31(1). 23–73.
Sitaridou, Ioanna. 2016. Reframing the phylogeny of Asia Minor Greek: The view from Pontic Greek. *CHS Research Bulletin* 4(1). 1–17.
Tsimpli, Ianthi Maria & Maria Dimitrakopoulou. 2007. The interpretability hypothesis: Evidence from *wh*-interrogatives in second language acquisition. *Second Language Research* 23(2). 215–242.
Tsimpli, Ianthi Maria & Maria Mastropavlou. 2007. Feature interpretability in L2 acquisition and SLI: Greek clitics and determiners. In Helen Goodluck, Juana Liceras & Helmut Zobl (eds.), *The role of formal features in second language acquisition*, 143–183. New York: Routledge.
van der Wal, Jenekke. 2016. Diagnosing focus. *Studies in Language* 40(2). 259–301.

Textual sources

Machairas Leontios, ΕΞΗΓΗΣΙΣ τῆς γλυκείας χώρας Κύπρου, ἡ ποία λέγεται Κρόνακα τουτέστιν Χρονικ(όν) [Recital concerning the Sweet Land of Cyprus entitled Chronicle]. http://users.uoa.gr/~nektar/history/2romanity/makhairas_chronicle.htm.
New Testament. http://www.ntgateway.com.
The Bible in Turkish. http://worldbibles.org/language_detail/eng/tur/Turkish.
Ἄχτι Τζετὶτ Καινὴ Διαθήκη [New Testament in Karamanlidika], 1838. https://anemi.lib.uoc.gr/metadata/4/5/8/metadata-39-0000472.tkl.

Christiane Bulut
Word order in Iran-Turkic

Abstract: Turkic languages are basically SOV, although their case morphology allows for a great deal of flexibility in the arrangement of sentence constituents. The present paper looks at word order properties in varieties of Turkic spoken in Central Iran, considering also the correlation between elements of the information structure, such as topic and focus, and suprasegmental intonation patterns, such as pitch and intensity contours. A striking feature is the frequency of the placing of 'Targets' in postverbal position, which seems to point to Iranian influence, or to a broader areal development in progress.

Keywords: language contact-induced developments; placing of 'Targets'; suprasegmental intonation patterns and information structure; Turkic word order properties

1 Introduction

This article deals with word order properties of Iran-Turkic, considering both genuine Turkic features and copied structures imitating Iranian models.

The examples for this article are taken from recordings of samples of free speech or 'texts' made in the region of Bayâdestân in Central Iran during three different field trips in the 1990s.[1] As the villages in this region are predominantly Turkic-speaking, the local varieties give a good impression of characteristic linguistic features and still ongoing developments in the southern subgroup of the Southwestern or Oghuz branch of Turkic.[2]

Working with texts (and not with isolated examples elicited with the help of questionnaires) makes it easier to trace discourse-related changes of word order properties, such as reference forms and topic or focus positions, in context. At the same time, the use of sound recordings offers the possibility to double-check the relationship between elements of the sentence emphasized by word order and their correlation with intonation patterns, such as suprasegmental or pitch accent.

---

[1] See Bulut (in print); references to examples from these texts are given as (no. of text: page number).
[2] For a classification of the Turkic varieties spoken in Iran see Anonby et al. (2020).

Christiane Bulut: Department of Turkish and Middle Eastern Studies, University of Cyprus, 75, Kallipoleos Ave., P. O. Box 20 537, 1678 Nicosia, Cyprus, E-Mail: bulut@ucy.ac.cy

https://doi.org/10.1515/9783110790368-008

2 Typological features of Turkic languages

With regard to their structural properties, Turkic languages are rather conservative; even geographically distant representatives of this genetically related group of languages are very similar to each other. Like most neighboring Iranian languages, Turkic languages are basically SOV, representing the subtype of 'rigid SOV' languages (Greenberg 1963: 63). They are strictly left-branching and, consequently, use postpositional constructions (some of which combine with additional case marking on the nucleus) instead of pre- or circumpositional phrases (Greenberg 1963: 71). As a rule, the defining element precedes the defined. Verbal and nominal stems combine with suffixes, i.e., Turkic represents the agglutinative language type. Verbal categories are consistently expressed by suffixes. Due to this agglutinative structure, there are almost no periphrastic verbal constructions.[3]

Genuine Turkic simple and complex sentences contain just one finite verb form, a verb fully marked for tense/mood/aspect and actant. Subclausal relations are frequently expressed via special non-finite verb forms ('subjunctors', e.g., nominalized verb forms/verbal nouns, and converbs/gerunds).[4] Essentially, *adverbial action clauses* are based on gerunds, while nominalized verb forms such as verbal nouns/derived 'deverbal nouns' and participles form *agent clauses* (relative clauses) or *nominalized action clauses* (complement clauses); see Johanson (1990: 199f.). They are integrated into the matrix clause via case morphology or postpositions. The relationship can facultatively be emphasized by a conjunction, which is usually a copy[5] from a contact language (e.g., conditional *agar* 'if' from Iranian).

3 Canonical word order in Turkic languages

Within the simple Turkic sentence, the unmarked word order is SOV (subject, object, predicate), with the verb in sentence-final position. While the nominative subject does not display morphological marking, a specific or definite direct

---

3 In contrast to Turkic languages of e.g., the Northwestern group, combinations of semantic verb and auxiliary verbs expressing actionality are rarely used in the Southwestern or Oghuz branch. Nor do they belong to the inventory of Iran-Turkic, although some few patterns imitating Iranian models appear across certain varieties.
4 The basic concept for the analysis of the Turkic sentence relies on Ergin (1958).
5 On the code copying model see Johanson (1992).

object is differentially marked with the accusative case (DOM). Items in the dative or directive case function as indirect objects, including semantic roles that are called 'Target' in this volume (see Asadpour, this volume); these are printed bold in the examples. The ablative indicates the source or the starting point of an action as well as a movement away from a person or object, among other functions. The locative case indicates the location.

The subject can be overtly represented by a noun or personal pronoun on the sentence surface; alternatively, in a construction which is called 'hidden subject' in Turkish, it is only marked by the personal suffix on the verb, the resulting pattern being (S)OV. As complements of the verb are unambiguously indicated by case marking, the order of constituents is quite flexible. One exception is the (morphologically unmarked) indefinite and non-specific direct object ('unmarked accusative', indicated by 'Ø'), which occupies the position directly preceding the verb.

The unmarked (S)OV word order often conveys all-new information in discourse, as is the case with new topics mentioned at the beginning of a text, or in independent statements. Ex. (1)[6] below displays a simple sentence containing new information in unmarked word order. The subject position is stressed by the use of an additional analytic pronoun *män* ('I, and no one else!'); the unmarked object takes its canonical position directly in front of the verb, while the Target is placed between the subject and the direct object, with the resulting word order being STØV:

(1) *män* **bu** *æmmə* *yizəmın* ***bašina***
 I DEM uncle daughter:POSS1SG.GEN head:POSS3SG.DAT
 bır *üš* *piyalə* *su* *dököm.*
 one three vessel water pour:VOL1SG
 'I will rinse the head of my cousin with three vessels of water.' (lit. 'I will pour three vessels of water on the head of my uncle's daughter.') [147: 62]

Adverbs or adverbial phrases can take any position in the sentence. Especially in the more official register of the language as spoken by bilingual male speakers, there seems to be no limit to the number of adverbs that can be added to a sentence.

Ex. (2) below contains three adverbs and one adverbial phrase in the locative case. The underlying main clause is based on the impersonal modal expres-

---

[6] The Turkic examples are given in a Turcological transcription, while IPA is used for the annotation of spectrograms. Note that the spectrograms are based on field recordings and therefore less precise than comparable items created from recordings made in a studio.

sion *gäräk* 'it is necessary.' Theoretically, any of the three adverbs could have been placed in front of the subject *ušaxlar*, but the initial position of the subject additionally stresses its topic function.

(2) *ušaxlar ævvæl dær hal-ə haːzır mædıræsädä bı'l-iğbaːr*
child:PL first in situation-EZ present school:LOC absolutely
gäräk faːrsı sohbæt elıyelär ... [27: 1]
necessary Persian speak:OPT3PL
'Now, first of all, it is absolutely necessary that the children speak Persian at school, ...'

4 Iran-Turkic

Iran-Turkic varieties display far-reaching contact-influenced changes in their lexicon, phonology, and syntax, while morphology is relatively conservative. In noun phrases, the Turkic-type left-branching possessive construction (as, for instance, in *æmmə yizəm-ın baš-i* in ex. 1, or *gærdænbændı-nıy muncuxlar-ı* in ex. 4) is often replaced by right-branching Iranian *ezâfe* constructions, where the so-called *ezâfe* particle links attributes to their headword. In addition, especially younger bilingual speakers tend to use prepositional Iranian constructions combined with Turkic flagging (case morphology and postpositions); see constructions such as: *æz bahräyn-nän* [from.PREP-Bahrain:ABL] 'from Bahrain', which displays both a copied preposition and a Turkic case marker. Similarly, examples such as *bæraye mısal ičü(n)* [for.PREP example for:POSP] 'for the sake of example', and *ba xatır-e kilase gedmäge ičü(n)* [with.PREP reason-EZ class:DAT go:VN.POSS3SG for:POSP] 'for the sake of (his) going to school' display both a copied preposition and a postposition, or a postposition plus case. In such combinations, the copied preposition is more or less a decorative element, as the morphological integration of the noun phrase into the matrix clause relies on Turkic morpho-syntactic elements, namely case marking and postpositional flagging. This double coding via genuine Turkic case morphology and/or postposition in addition to copied Iranian prepositions is syntactically redundant. Possibly, it is indicative of some development in progress, which may eventually lead to a replacement of Turkic case marking and postpositions by Iranian-type prepositional phrases.

Pre- or postpositional phrases as well as adverbs and adverbial phrases can appear anywhere in the sentence. Ex. (3) illustrates the high frequency of copied prepositional phrases and *ezâfe* complexes (in bold), all of which reverse the

Turkic order of constituents in noun phrases. This example is characteristic of the formal speech of bilingual speakers.

(3) *Væli* **zæbå:n-e** *Türki* æz næzær-e ædæbiyat čox
but language-EZ Turkic from point-EZ literature much
yæni-tär-di **nisbät be zæbå:n-e Fa:rsi.** (...) Äyär
rich:COMP.COP3SG regard to language-EZ Persian if
Ærabüni, **xod-e kälæmå:t-e Türki** ke **zæbå:n-e**
Arabic:ACC own-EZ words-EZ Turki CONJ language-EZ
Fa:rsi *ičindä* var götüräy, æslæn
Persian inside (POSP) exists take_away:OPT2SG actually
zæbå:n olmaz bilæsinä diyey. [29: 11–13]
language be:NEG.AOR3SG itself:DAT say:OPT2SG
'Yet, with regard to its literary production, the Turkic language is much richer, compared to Persian. (...) If you take away all the Arabic, all these Turkic or Arabic words that are part of Persian, there would actually remain nothing you could call a language.'

Under the influence of structurally different languages, Turkic varieties have almost completely restructured the syntax of complex sentences according to Indo-European (mostly Iranian) and Semitic models (see Bittner 1900). Right-branching dependent clauses (relative, complement, or adverbial clauses) based on finite verb forms replace left-branching constructions based on non-finite subjunctors. Additionally, the dependent clause may be connected to the main clause by a copied *conjunctor*, which, at the same time, expresses the semantic relation between main clause and dependent clause.

4.1 Prepositional flagging and right-branching relative constructions

In theory, such a typological change in clause-combining strategies would imply that Iran-Turkic also developed different alignment patterns in noun phrases. Yet, *ezâfe* complexes (which violate the Turkic principle of possessor → possessum, or defining → defined) and copies of Iranian-type prepositional phrases still establish alternative structures which are used along with genuine Turkic patterns. The need to anchor prepositional phrases in the basic code via additional Turkic case morphology or postpositions also shows that they are still not fully functional elements of Turkic syntax. In its present shape, Iran-Turkic has reached an in-between stage, displaying copied right-branching structures in the syntax of dependent clauses, as opposed to predominantly left-branching noun phrases and postpositional constructions. In ex. (4) below, the right-branching relative clause follows its basic segment/head, which violates the Turkic rule of defining (= RC) → defined (= head/basic segment).

Although the adoption of Iranian prepositional flagging, which changes the position of flagging from the end to the beginning of a phrase, could motivate postverbal placement according to distance principles, postverbal elements usually preserve their Turkic head-final structure. In fact, ex. (4) displays the exact opposite word order than predicted. The object phrase includes an Iranian-style relative clause that is postposed, thereby increasing the distance between the accusative case marker and the verb, and the Target appears in postverbal position despite head-final flagging. This suggests that other factors also play a role in the placement of constituents with respect to the verb.

(4) o gærdænbændɪnıy muncuxlarını kı geğä
 that necklace:POSS3SG.GEN beads:POSS3PL.ACC CONJ at_night
 hær nä tabmıš‿ımıš qæræŋguluxda götürär
 as_much find:EV.PF3SG darkness:LOC take_away:PRS3SG
 gætürär evä. [144: 24]
 bring:PRS3SG home:DAT

'She collects as many of the beads of her necklace as she is able to find in the dark of night, and takes them home.'

The Turkic strategy of forming adverbial clauses based on gerunds also seems to be largely extinct across Iran-Turkic varieties. A lone survivor is the gerund in {-(y)ElI} 'since, as long as', which occurs sporadically in the texts I collected from speakers in their eighties. Yet, it was a rare pattern that even these speakers used only as an alternative structure along with Iranian models in forming temporal clauses. Younger speakers no longer apply gerundial syntax. Consequently, most temporal clauses imitate Iranian models; see *o væx ke bɪz yå:d verärde:k^h* 'ever since we can remember' in ex. (5b) below, where the Iranian temporal clause is actually a relative clause on the basic segment/head (*vaxt* 'time'). Note that the same speaker also uses a construction based on the gerund in {-(y)ElI} to form temporal clauses.

(5) a. *Yad veräli čirax dasti var.*
 remember:GER lantern existing
 'As long as I remember/one remembers, there have been lanterns ... '

Word order in Turkic-type dependent clauses based on non-finite verb forms (gerunds and verbal nouns) is much more restrictive than in 'main clauses' or simple sentences based on a single finite verb form. The non-finite subjunctors always take clause-final position; they mark the border between dependent clause and main clause. Consequently, constituents of dependent clauses will not appear in postverbal position, that is, after the gerund or verbal noun; see ex. (5a). Following the basic rule that defining elements precede the defined, Turkic dependent clauses based on non-finite verb forms precede the main clause containing the finite verb.

4.2 Topic, focus, and intonation patterns

In samples of spoken utterances, the semantic function of sentence constituents is additionally emphasized by stress patterns. Turkic languages display two kinds of stress, namely, word accent and suprasegmental stress or pitch[7] ('sentence melody').

Word order is one strategy to mark the function of elements in discourse (topic, focus, etc.), and it often correlates with suprasegmental stress patterns. In Iran-Turkic, though, the picture is much more complicated. Firstly, the range of the pitch contour is gender-specific and varies across different age groups. It is extremely high in monolingual female speakers – the high-pitched intonation and the oscillating modulation creating the impression that they are 'singing' their way through the sentence. Older male speakers display a less differentiated pitch contour, with a much lower basic frequency. With bilingual male speakers and representatives of the younger generations, the pitch contour is even less varied – which may point to the influence of imitated Iranian (predominantly Persian) intonation models. To further complicate the picture, speakers of Iran-Turkic varieties use word order properties and syntactic patterns copied from their (mostly Iranian) contact languages along with the original Turkic strategies.

In unmarked word order (SOV), the topic is often identical to the subject appearing in sentence initial position, while the position immediately preceding the verb is the focus position, containing the central elements of information. Essentially, focus (the most important item of the rheme or comment) coincides with pitch accent in suprasegmental intonation patterns, or the highest peak of the sentence melody. The topic or theme, or the element the sentence refers to, also receives intonational stress; in complex sentences, it may be set off from the rest of the sentence by an intonational pause. Both pitch and intensity usually fall after the focused element in preverbal position, while the peak of the pitch con-

---

[7] Pitch is actually a perceptional concept. Pitch contours depict the oscillation of the vocal folds at the level of F0 (fundamental frequency). High frequency is perceived as high pitch.

tour may also include the first syllable of the verb – which is mostly identical to the verb stem. Consequently, elements in postverbal position do not receive suprasegmental stress, while intensity also decreases (see the red line in the figures below). In complex sentences according to the Iranian model, the last syllable of a sentence-final verb or of the last element in postverbal position is lengthened and shows a rising pitch contour (the so-called drawl, indicated in the examples by ↗), if a dependent clause is going to follow; see example (5b) below:

(5) b. o   væx' ke   bız   yå:d   verärde:kʰ ↗,
        that time CONJ we   remember:AOR.PST1PL
        e:l   gælä:rdə   **bızım bu   daylaræ:**   [3: 1]
        tribe come:AOR.PST3SG our   DEM mountain:PL.DAT
        'Ever since we can remember, nomads used to come to our mountains, ... '

In cases where the subject is unambiguously marked by personal suffixes of the verb (as is normally the case with 1st and 2nd persons), an overt marking by means of analytic pronouns implies additional stress or emphasis on the subject position (see ex. 1 above, and also the first sentence of ex. 10 below).

Spectogram of ex. (1):

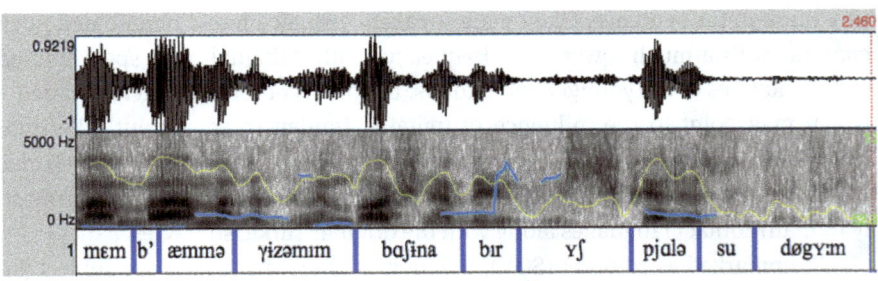

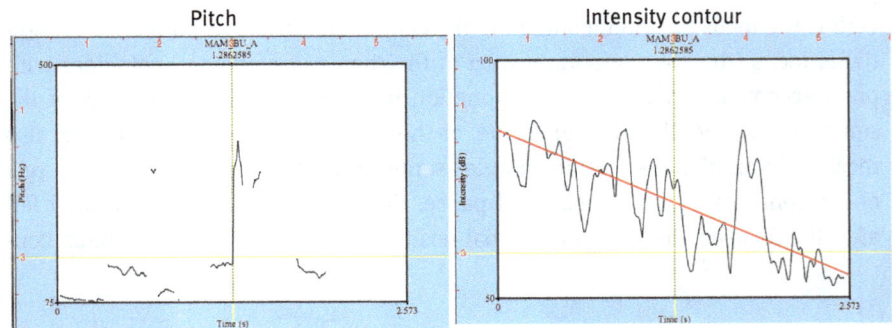

The spectrogram of ex. (1) above shows high intensity of the pronominal subject *män* 'I', and also of the complex Target *bu æmmə yizəmɪn bašɪna* 'on the head of my uncle's daughter' in preverbal position. Falling intensity and pitch contours mark the end of the utterance, with the direct object and the verb receiving low stress (almost reduced to a whisper); this is characteristic of the suprasegmental intonation patterns of older speakers.

The subject may also be moved to the focus position before the verb. In this way, the direct object appears in sentence-initial position (OSV); that is, the topic position at the beginning of an utterance containing new information. Ex. (6) below illustrates the classical topic → focus alignment. The speaker introduces a new topic, namely the weaving of rugs. He emphasizes that his family used to be involved in the process of producing rugs by putting the reflexive pronoun *özümüz* '(we) ourselves' in the focus position immediately before the verb. Intensity and pitch contours are highest at the beginning of the sentence, declining continuously toward the end; consequently, stress and intensity are concentrated on the sentence-initial direct object in topic position:

(6)　o　　qa:linə　bɪz　özümüz　　toxomašuq!　　　　　　　　　[54: 1]
　　　that rug:ACC　we　ourselves　weave:PF1PL
　　　'This rug we have woven *ourselves*!'

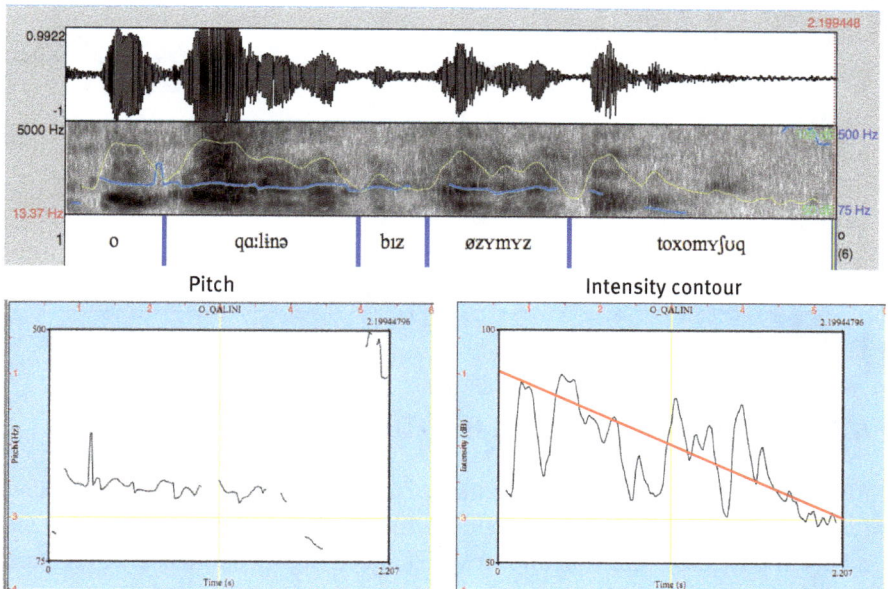

The Target (T) or goal in the dative may be placed in preverbal position, as in STØV in ex. (1) above and in STV in the second half of ex. (7) below. The position of T (*bu türkiyä ma:šmnärmə*) immediately before the verb coincides with the usual rise in the pitch contour and stress on the focus position; compare also the pitch/intensity contours in the annotated spectrogram of the second half of the sentence below.

(7)   hæmədænnän   kɪ   gælɪrde:m ↗,   män
     Hamedân:ABL   CONJ   come.IMPF.1SG   I
     bu   **türkiyä**   **ma:šmnärmə**   mɪnde:m ↗   [28: 1]
     DEM   Turkey   car:POSS3PL.DAT   board:PST1SG
    'When I was on the way back from Hamedân, I took (lit. boarded) one of these Turkish overland buses; ... '

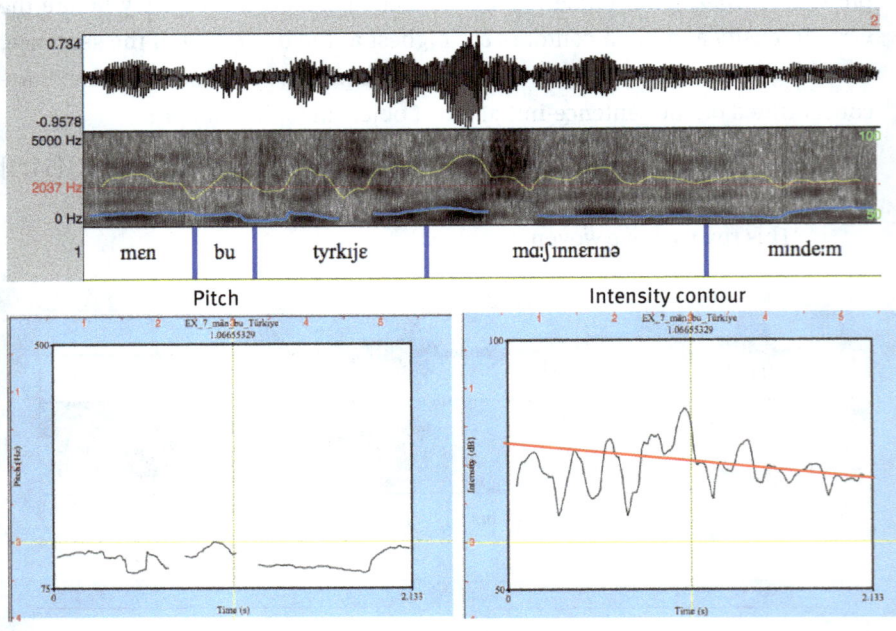

4.3 Postverbal elements in Iran-Turkic

Although word order in Turkic languages is dominantly verb-final, it also allows for constituents in postverbal position. In Iran-Turkic varieties, though, elements in postverbal position appear with a much higher frequency. Particularly Targets that are not in focus position are placed in the (unstressed) postverbal position; see example (8) below:

(8) tæmam gedıllär **šærısdanå**. [13: 33]
    all   go:PRS3PL   town:DAT
    ' ... , they all move into town.'

In some instances, the direct object appears in postverbal position (S)VO, which coincides with the lowest level of pitch; see ex. (9) below. Note that the verb *bırax-* 'to leave (behind)' represents the semantic group of change-of-state or inchoative verbs.

(9) bıraxıllar   bu   kändısdanɨ [13: 33]
    leave:PRS3PL   DEM   countryside:ACC
    'people (lit. they) leave the countryside,'

In a few instances, the unmarked object is also found in postverbal position; see the first sentence with the subject in topic position (SVØ) in ex. (10) below. Pitch and intensity decline after the verb stem *tox-*; consequently, the unmarked object in sentence-final position coincides with the lowest point in the pitch contour. In the subsequent sentence, the unmarked object *kilim* is back to its canonical position.

(10) næ,   män   özüm   toxomazdəm   xalɨ,
     no   I   myself   weave:NEG.AOR PST.1SG   rug
     kılım   toxomıšæm, ... [54:2]
     kelim   weave:PF1SG
     'No, I wasn't into rug weaving **myself**; I have made kelims, ...'

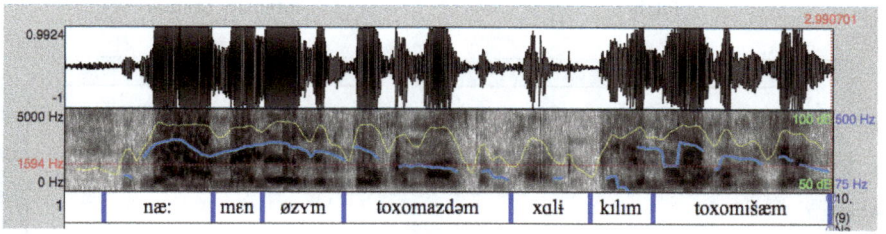

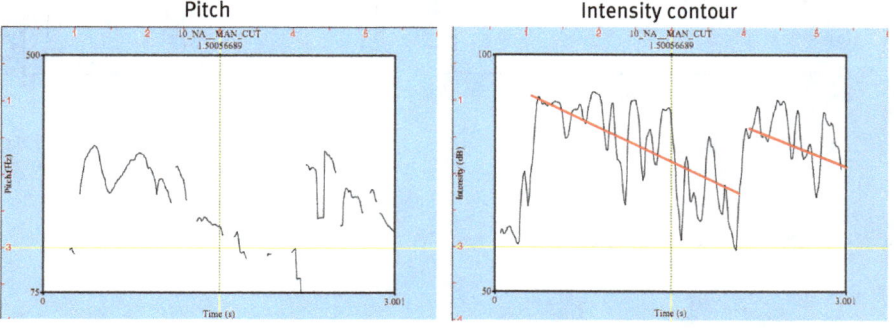

Very rarely, the predicate noun is placed in postverbal position, which may occur in connection with the auxiliary *ol-* 'to become'; see ex. (11) below. In an iconic way, this arrangement of auxiliary → predicate noun seems to stress the inchoative semantics of the auxiliary, the process leading to the result. As pitch and intensity contours fall after the verbal element (in this case the auxiliary 'to become'), in both sentences of ex. 11 the predicate noun is in an unstressed position; see the spectrogram below:

(11) *gælırdım   bädænum   olurdu   yåra,*
      come:IMPF1SG   body:POSS1SG   become.IMPF3SG   wound
      başim   olurdu   sırkä,   bi:tʰ.   [70: 4]
      head:POSS1SG   become:IMPF3SG dandruff   lice

'And when I returned (from the public bath), I would break out in a rash, my head would become (infested with) dandruff and lice.'

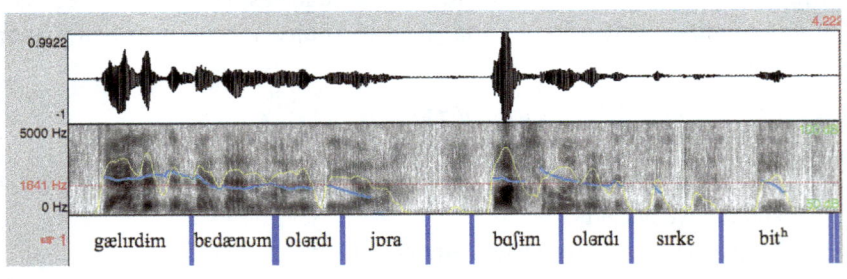

Pitch   Intensity contour

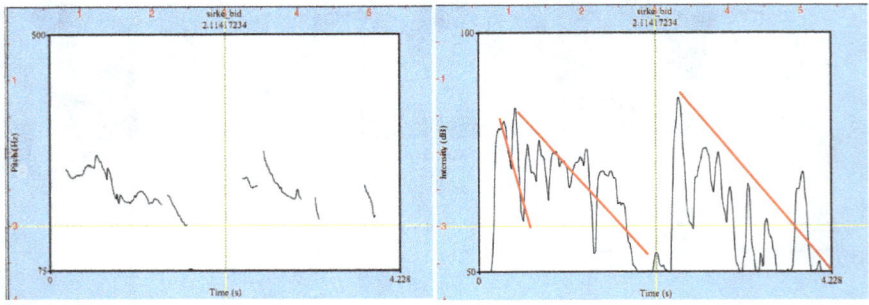

4.4 (S)OV-T or (S)ØV-T as unmarked word order

In simple sentences containing both a direct object (marked or unmarked) and a Target, the canonical word order in Iran-Turkic seems to be (S)OV-T, or (S)ØV-T, respectively. As all elements of the examples below contain new information, this pattern obviously represents the unmarked word order. As usual, the pitch and intensity contours fall sharply after the verb, necessarily placing the T in an unstressed position; see examples (12), (13), (14), (15), and (16) below.

(12) Boydanɨ   säpæ:rdɨq   **quru   torpaya:**   [46: 8]
       wheat   scatter:AOR PST1PL   dry   ground:DAT
       'We usually sowed the wheat on the dry ground.'

(13) gänä   o   börk^hɨ   giyäräm   **bašɨma,**
       again that   hat:ACC   put_on:AOR1SG   head:POSS1SG.DAT
       ælğäklärə   giyäräm   **ælɨmə.**   [52: 23]
       glove:PL.ACC   put_on:AOR1SG   hand:POSS1SG.DAT
       'I put this hat on my head again, and I put the gloves on my hands.'

(14) bu   bʊlɨyə   aldɨ   atde   **dæŋyäyä.**   [147: 14]
       DEM   fish:ACC   take:PST3SG   throw:PST3SG   sea:DAT
       'He took the fish and threw it (back) into the sea.'

Essentially, the same word order pattern also appears with unmarked objects; see ex. (15) below, which illustrates SV, followed by (S)ØVT.

(15) bo: oylannær   getdəlär   šal   sallærdəlær
       DEM   boy:PL   go:PST3PL   cummerbund   let_down:AOR PST3PL
       bo   æmmɨsɨnɨn   pağasɨnæ.   [147: 43]
       DEM   uncle:POSS3SG.GEN   chimney:POSS3SG.DAT
       'These boys went along and let their cummerbunds down their uncle's chimney.'

In ex. (16) below, the postpositional phrase æl-enæn 'by hand' follows the postverbal dative. It is pronounced like an afterthought, with very low pitch and set off from the sentence by a long intonational pause.

(16) toxum   boyda   säpä:rdik   **buna:,**   æl-enæn.   [46: 6]
       seed   wheat   scatter:AOR PST1PL   this:DAT   hand:POSP
       'We sowed seeds, wheat, on it – by hand.'

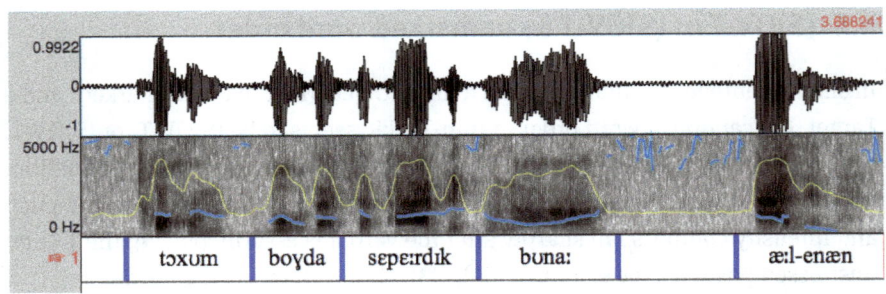

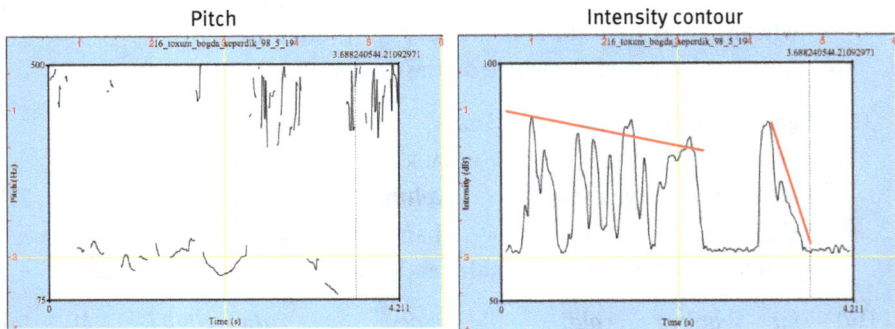

4.5 Motivation for postverbal placement

In many instances, deviations from the unmarked word order have a pragmatic function; they are discourse-oriented, or in other words, related to the information structure of the context in which the utterance appears. Information already mentioned in previous discourse often establishes the topic/theme of the utterance, while the focus is the main element of new information (comment/rheme) on this topic.

Iran-Turkic displays strong tendencies to move constituents of the simple sentence that directly depend on the verb (verbal complements/arguments), from a preverbal or even focus position to an unstressed postverbal position. The most striking feature is the frequent postverbal placement of the indirect object denoting the Target or goal of a movement/action – which, from the perspective of both word order properties and the accompanying pitch contours, seems to downgrade the value of the information these elements convey.

Across Turkic languages, objects in the dative or accusative case may sporadically appear in postverbal positions. Iran-Turkic, on the other hand, uses postverbal placement of goals/Targets with a much higher frequency than preverbal placement. While in STV (as in ex. 7 above) T is clearly in focus position,

SVT seems to be the most common pattern for unmarked word order; see ex. (5) and (8) above. In sentences containing both a direct object and a Target, the most frequent word order is (S)OV-T; see ex. (4), (12), (13), (14), (15), and (16) above. However, STOV also occurs in ex. 1

There may be several reasons for this change of word order properties. The pattern (S)OV-T is probably motivated by the formal need to keep the ACC closer to the verb, or in the stressed position preceding the verb. This would imply a hierarchy of constituents, where the ACC ranks higher than the DAT/Target. Another reason for (S)V-T or (S)OV-T could be iconicity; the postverbal position of the Target after verbs indicating motion seems to underline the dynamics of the verb semantics: motion (first) leading to the goal/Target (second). In both cases, T in postverbal position contains new or necessary information; nonetheless, it has been transferred from the preverbal/stressed position to the falling pitch contour after the verb. Consequently, semantic values or discourse pragmatics, such as topicalization or focusing, do not seem to substantially alter the characteristic suprasegmental intonation patterns of Turkic, where the pitch contours fall after the verb.

Another explanation for (S)VT or (S)OV-T word order patterns is the structural influence of languages such as Modern Persian or Kurdish, where the goal tends to be only marked by position, and not by case morphology. In some Iranian contact languages, the Target is defined by its canonical postverbal position. Kurdish, for instance, marks the Target morphologically by oblique case (indicating genitive, dative, or accusative case). This oblique form is defined as directive or Target by word order properties, namely the placement in postverbal position; see, for instance, SØV-T in: *ez pere dıdım te* [I:DIR money give:PRS.IND1SG you:SG.OBL] 'I give you money,' or SVT in: *kurê mın dıçe mektebê* [son:EZ.M I.OBL go:PRS.IND3SG school:OBL.F] 'My son goes to school.' For Turkic, where the morpho-syntactic function of the indirect object is explicitly specified by dative case marking, this kind of double coding seems redundant. In Modern Persian, another potential contact language, the Target is placed in preverbal position and should be (but is not always) introduced by a preposition (e. g. *be* 'to' etc.); compare PREP-TV in: *be maktab mîravam* [to.PREP school go:PRS.IND1SG] 'I go to school.' Instead of using prepositional phrases in preverbal position, colloquial Persian may also put the morphologically unmarked Target in postverbal position to indicate directive case, as in: *mîram maktab* [go:PRS.IND1SG school] 'I go to school.' Consequently, the increased tendency in Iran-Turkic to place the Target in postverbal position[8] could be interpreted as a copy of structural features from the Iranian contact lan-

---

[8] This development can also be observed in Turkic varieties of adjacent regions; see Bulut (1999).

guages, or as an areal feature – a shared feature developing across structurally different languages of the region.

Scholars have also attempted to explain word order in formal terms, for instance assuming that the verb and the heads of its phrases should be minimally distant from one another (see Wasow, this volume, for details). Strictly left-branching languages like Turkic are assumed to place longer phrases before shorter ones. This does not seem to be the case in Iran-Turkic; see, for instance, ex. (15), where a rather complex head-final Target follows the verb instead of preceding it.

5 Historical perspective and outlook

Across the larger area comprising Iran, present-day Iraq, and adjacent regions, Turkic languages display a great deal of contact-induced changes, especially on the levels of lexicon, phonology and syntax. This is partly the result of a development set in motion nearly a millennium ago, when large numbers of Turkic-speaking populations under the leadership of the Saljuqs entered the central lands of the Eastern Caliphate. Becoming integrated in the Islamic cultural sphere, Turkic adopted a large amount of Islamic (that is, Arabic) lexicon and terminology, mostly via Persian. While Arabic was the domain of theology, law, and the natural sciences, Modern Persian was adopted as the official language and the language of administration and literature in a number of Turkic-Islamic states in Central Asia, greater Iran, and Anatolia. Turkic, on the other hand, served as the language of communication among different ethnic components of the great tribal confederations, and, consequently, the military. This led to the emergence of a high rate of bi- or multilingual speakers, furthering contacts and the exchange of structural features between different languages – especially those belonging to the Turkic and Iranian language families. Due to its function as a pan-regional means of communication (a spoken lingua franca) and the relatively low prestige of this language, Turkic has been extremely open to influences from structurally different contact languages.[9]

The Turkic varieties in contact with Iranian in the broader area including Anatolia, Iran, present-day Iraq, and parts of Central Asia have copied numerous morpho-syntactic elements and complex phrase structures of predominantly

---

9 For more information on the historical development of Iran-Turkic and the South Oghuz dialect group, see Bulut (2014), and Johanson and Bulut (2006).

Iranian origin which violate basic rules of Turkic word order. These processes of structural copying have already created syntactic constructions and word order properties that are diametrically opposed to the typological features of Turkic.

Under the prevailing conditions in Iran, where the educational system solely promotes the national language, the influence of Modern Persian on Turkic and other minority languages has been constantly growing. This may lead to an intensification or acceleration of the development of contact-induced structural changes in the affected languages, which should be closely followed and documented.

Abbreviations

1, 2, 3	1st, 2nd, 3rd person
ABL	ablative
ACC	accusative
AOR	aorist
COMP	comparative
CONJ	conjunction
COP	copula
DAT	dative
DEM	demonstrative pronoun
DIR	directive
EV	evidential
EZ	ezafe particle
F	feminine
GEN	genitive
GER	gerund
IMPF	imperfective
IND	indicative
LOC	locative
M	masculine
NEG	negation/negative
OBL	oblique feminine
OPT	optative (mood)
PF	perfect
PL	plural
POSP	postposition
POSS	possessive
PREP	preposition
PRS	present tense
PST	simple past
SG	singular
VN	verbal noun
VOL	voluntative (mood)

References

Anonby, Erik et al. 2020. Turkic languages in the Atlas of the Languages of Iran (ALI). *Turkic Languages* 24(2). 290–308.
Asadpour, Hiwa. this volume. Word order in Mukri Kurdish – the case of incorporated targets.
Bittner, Maximilian. 1900. Der Einfluß des Arabischen und Persischen auf das Türkische. *Sitzungsberichte der philosophisch-historischen Classe der kaiserlichen Akademie der Wissenschaften* 142. 38–119.
Bulut, Christiane. 1999. Klassifikatorische Merkmale des Iraktürkischen. *Orientalia Suecana* 48. 5–27.
Bulut, Christiane. 2014. Turkic varieties in West Iran and Iraq. Representatives of a South Oghuz dialect group? In Heidi Stein (ed.), *Turkic in Iran. Past and present*, 15–99. Wiesbaden: Harrassowitz.
Bulut, Christiane. in print. *The Turks of Central Iran. Language and culture in the Turkic region of Bayâdestân.* Vol. 1: Texts & Vol. 2: Commentary. Wiesbaden: Harrassowitz (Habilitationsschrift Mainz, Nov. 2006).
Ergin, Muharrem. 1958. *Türk Dil Bilgisi.* İstanbul: Boğaziçi Yayınları.
Greenberg, Joseph. 1963. Some universals of grammar with particular reference to the order of meaningful elements. In Joseph Greenberg (ed.), *Universals of language*, 73–113. London: MIT Press.
Johanson, Lars. 1990. Subjektlose Sätze im Türkischen. In Bernt Brendemoen (ed.), *Altaica Osloensia. Proceedings from the 32nd meeting of the Permanent International Altaistic Conference* 193–218. Oslo: Universitetsforlaget.
Johanson, Lars. 1992. *Strukturelle Faktoren in türkischen Sprachkontakten* [= Sitzungsberichte der Wissenschaftlichen Gesellschaft an der Johann Wolfgang Goethe-Universität Frankfurt am Main 29/5].
Johanson, Lars & Christiane Bulut (eds.). 2006. *Turkic-Iranian contact areas. Historical and linguistic aspects.* Wiesbaden: Harrassowitz.
Wasow, Thomas. this volume. Factors influencing word ordering.

Valeriya Lemskaya
Word order variation in Chulym Turkic of Siberia

Abstract: Despite the attention by Turkologists since ca. 1940s, Chulym Turkic remains understudied, especially its morphosyntax. This article focuses on the word order in Chulym Turkic simple sentences. The author's own field data of Lower Chulym Turkic (gathered 2007–2010 with the last known competent speaker) are compared to those collected in the 19th–20th centuries by W. Radloff and A. Dulzon. Middle Chulym Turkic field data (of 2006–2021) are compared to those earlier recorded by S. Malov, M. Abdrahmanov and R. Biryukovich. Conclusions on word order variation among the Chulym Turkic dialects and their possible emergence due to language contact are made.

Keywords: Chulym Turkic; language contact; (nearly) extinct language; syntax; word order

1 Chulym Turkic in the language contact zone

Chulym Turkic is a critically endangered "transitory" South Siberian Turkic language, which is part of the Border Turkic group along with the Yenisey Turkic (Khakas and Shor) and Sayan Turkic (Tuvan and Tofa) (Schönig 1997: 125). The ethnic group of Chulym Turks is rather a small one with 355 people counted during the 2010 All-Russian National Census. They inhabit remote Siberian villages in two regional provinces of the Russian Federation, mainly Teguldet (the center of the Teguldet District, Tomsk Region) and Pasechnoye (the Tyukhtet District, Krasnoyarsk Territory). Both settlements either stand on the bank of, or are in quite close proximity to the river Chulym, the one that has given name to the local population (folk etymology goes as far as to consider it a variation of *čul-(I)m* river-1SG.POSS 'my river'). There have been other designations for them used in the Russian tradition, like *isašnye* (< 'yasak (i.e., fur tax)-paying people'), *čulymskie tatary* 'Chulym Tatars' and even *čulymskie xakasy* 'Chulym Khakas'; see the discussion in (Lemskaya 2016). To date, the official

---

Valeriya Lemskaya: Tomsk State Pedagogical University, Kievskaya, 60, Tomsk, Russia 634061 and Tomsk State University, Lenina 36, Tomsk, Russia 634050, E-Mail: lemskaya@gmail.com

https://doi.org/10.1515/9783110790368-009

designation of this people in Russia is *čulymcy* 'Chulymites' and the language – *čulymskij jazyk* 'the Chulym language' (All-Russian census 2010).

Previously, two dialect groups were distinguished, each having subdialects: the now extinct Lower Chulym (Yeži, Yači, Čibi, Küärik and Ketsik) and the still existing Middle Chulym (Tutal and Melet) (Dulzon 1973). Each of the subdialect designations reflects the ethnic subgroups (with administrative *volost* divisions) who lived along the flow of the Chulym river and its tributaries. Historically, up to 14 such volosts were attested in the area (Georgi 1780), so more dialects of Chulym Turkic could have existed there. This topic, however, deserves a separate study.

Based on historical data by Zemlja (2001: 246), Boyarshinova (1950), Radloff (1989: 92), and Dulzon (1973: 16 etc.), it is possible to mark an approximate habitat of the Turks in the Chulym basin who can be considered Chulym Turks.

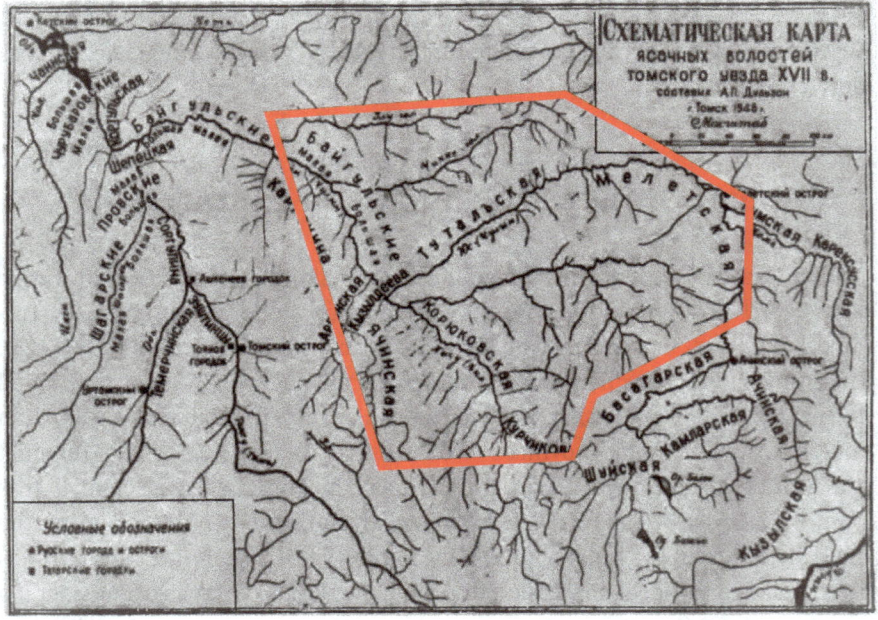

Figure 1: An outline map from (Dulzon 1952: 84) with a markup of the probable historical Chulym Turkic habitat.

The Turks – the Yenisei Kyrgyz – are believed to have started appearing in the north of the marked area (the Middle and Lower flow of the Chulym river) since as early as the 9th century (Belikova 2001: 86) or even the 7th–8th centuries (Dulzon 1973: 27). Migrations continued into the 12th–13th centuries, the newcomers mixed with the already residing local population and the ethnic group of Chulym Turks was formed by the 15th century (cf. Dulzon 1973: 27). However, Chulym Turks never had a systematized state system, nor any unification of the culture or language. W. Radloff also noted that South Siberian Turkic tribes lived quite intermixed (*"племена живут так перемешанно между собой"*; Radloff 1868: XIV).

Thus, Chulym Turkic can be considered a set of varieties of a (geographical) dialect *cluster* rather than a unified/homogeneous language; and the term *Chulym Turkic language* is rather an approximated one and is not to be considered signifying a single language.

The Turks around the Chulym River have always been in contact with various ethnic and linguistic groups, possibly at first with the Yenisseic (Paleosiberian) and Samoyed (Uralic) peoples. In addition, the Turkic groups that merged in the Chulym Turks, the Shor, and the Khakas were in constant contact with each other, and starting from the beginning of the 17th century, with adoption into the Russian State, they came into close contact with speakers of Russian (Belikova 2001; Dulzon 1973). It is with the founding of the Achinsk town that the Turks of the Chulym were divided and those who had settled up the river were later included into the Khakas people. Bilingualism has long been a common feature of the local population:

> Dagegen was von ihnen in den Dörffern zwischen denen Russen wohnet, ist meist alles getaufft, sprechen Rußisch, auch so daß man sie nicht unterscheiden kan. ["Those who live in the villages among Russians, almost all are baptized, speak Russian, and in such a way that it is impossible to distinguish them from Russians"] (Strahlenberg 1730: 279)

and

> "Теперь эти татары испытываютъ на себѣ вліяніе русскаго элемента и съ каждымъ годомъ русѣютъ. Болѣе всѣхъ другихъ родовъ сохранилъ въ себѣ татарщины Кизильскій родъ, живущій по верхнему теченію Чулыма, въ предѣлахъ нынѣшней Енисейской губерніи"
> ["Now these Tatars are experiencing the influence of the Russian element on themselves and every year they become more Russified. Among the other clans, the Kizil clan living upstream of Chulym within the present Yenisei Province has retained the Tatar features most"] (Kostrov 1828: 18).

Thus, the Chulym river basin has long been a place of constant contact between the speakers of different language families: Turkic (Chulym Turkic varieties), Paleosiberian (Ket, Yugh, Pumpokol), Uralic (Samoyed), and Indo-European (Russian).

2 Word order variation in Lower and Middle Chulym Turkic

Chulym Turkic is a moribund language with no established standard writing system or speech norm (cf. Pomorska 2004: 18). Linguists have recorded Chulym Turkic data in either Cyrillic (W. Radloff, A. P. Dulzon, M. A. Abdrahmanov, R. M. Biryukovich, R. A. Boni) or Latin transcription (A. P. Dulzon, R. M. Biryukovich, G. D. S. Anderson & K. D. Harrison, Y.-S. Li et al., V. Lemskaya). Archived and published texts can be classified as oral texts. The genres of the texts are either folklore texts (fairy tales, legends, folk rhymes) or everyday texts. There have been attempts by the speakers themselves to record stories and text samples in their mother tongue using the only alphabet they knew – Russian. Their writings were not systematic partly because the Russian alphabet does not reflect many Turkic sounds or phonemes. These texts are available for analysis but very scarce in number[1]. Before proceeding with analysis of language data, the characteristic of the language material itself will be given.

2.1 Lower and Middle Chulym Turkic speakers and corpus

The term *Chulym Turkic* today implies the Middle Chulym varieties only. The Lower Chulym dialect is considered extinct since 2011 with the death of the last known speaker. Lower Chulym speakers had been far fewer in number compared to the Middle Chulym group. The number of Middle Chulym Turkic fully fluent speakers today is not to exceed 10 with additionally a dozen (or maybe even two dozen) people who understand the language but do not speak it. This number is my personal estimation supported by colleagues (mostly Mikhail Popov, a postgraduate student at the Institute of Anthropology and Ethnography, Russian Academy of Sciences; p.c.). I estimate there are not more than 50 Chulym Turkic speakers combined; cf. Anderson and Harrison's (2006) estimation of only the Teguldet District's fluent speakers to be fewer than 20 with additional 20 to 30 semi-speakers remaining.

---

[1] It must be noted, however, that three volumes of Middle Chulym Turkic language materials prepared for the community have finally seen light recently: V. M. Gabov's translation of the Gospel of Mark into the Tutal subdialect under V. Lemskaya's editing was published in 2019 (the author's choice was the Russian alphabet with three additional graphemes), and A. F. Kondiyakov and V. Lemskaya's two volumes of the Melet subdialect's lexicon and annotated texts – in 2021 (the graphic system adopted was the Russian alphabet with seven additional graphemes).

The total number of annotated Chulym Turkic texts available for the current paper's analysis is 102. These texts contain 54,271 words in 7,013 sentences. The subdivision of the corpus is as follows:

Table 1: Chulym Turkic annotated text corpus.

(Sub)dialect	Texts	Source	Type/Genre	Words	Sentences
Lower Chulym					
Küärik	1	Radloff (1868)	legend	4,369	408
Yeži	2	TEV (1886)	religious; translation	210	20
Yeži	5	Dulzon (1952), (1966)	life story	545	74
Yeži	4	Lemskaya (2007–2010)	life story	463	111
Total				5587	613
Middle Chulym					
Tutal	2	TEV (1886)	religious; translation	185	20
Tutal	4	Malov (1909)	folk story	357	70
Tutal	2	Abdrahmanov (1970)	life story; legend	1,711	250
Tutal	19	Biryukovich (1971: T. III), 1984; Boni (1973)	life story; legend	1,793	422
Tutal	2	Lemskaya (2006–2018)	religious; translation	20,878	2,019
Melet	3	Biryukovich (1971: T. VI)	folk story	1,615	336
Melet	60	Lemskaya (2006–2018)	folk story; rhymed song; life story	10,761	1,533
Melet	11	Lemskaya (2006–2018)	religious; translation	5,080	508
Melet	1	Lemskaya (2006–2018)	folk story; translation	670	87
Tutal and Melet	4	Lemskaya (2006–2018); Tokmashev (2015)	life story; dialogue	5,634	1,155
Total				44112	6400

Most of the archived texts from the corpus have been annotated and published (Lemskaya 2010, 2012, 2013, 2015, 2017, 2020a). As of today, the texts I have collected and annotated have not been printed yet, and most of my audio (ca. 190 hours) and video (ca. 70 hours) recordings made between 2006 and 2021 among the Chulym Turkic speakers have not been analyzed and are sadly to be left out of my current analysis. A lot of my records (with sporadic analysis), however, are stored and can be accessed at the ELAR and Lingvodoc archives.[2]

2.2 Word order variation in Lower Chulym Turkic

As noted above, the limited number of Lower Chulym texts still allows analysis of both text and sentence structure in a diachronic perspective. In my publication (Lemskaya 2020b), I have tried to analyze the information structure of the Lower Chulym simple sentence from a diachronic perspective in the texts of the 19th–21st centuries.

For my current analysis, about 20 simple sentences were selected from each dialect's group of texts. Their word order was analyzed with regards to the subject, verb predicate, direct and indirect object, adverbial, etc. (cf. Comrie 1989). The following table containing a comparison of Lower Chulym Turkic word order patterns was composed:

Table 2: A comparison of Lower Chulym Turkic simple sentence word order types from a diachronic perspective.

Texts from 1868	Texts from 1886	Texts from 1946–1966	Texts from 2007–2010
Advl – S – O(id+d) – V	Advl – S – V		Advl – S – O(d) – V
O(d) – Advl – V	O(d) – Advl – O(id) – V		S – Advl – O(id) – V
S – Advl – V	Advl – O(d) – V		S – O(d) – Advl – V
S – V	S – V		S – O(d) – V
O(d) – V	O(d) – (O(id) –) V		Advl – S – V
Advl – V			S – Advl – V
			S – V
			O(d) – V
			Advl – V
	Advl – V – O(d)	S – Advl – V – Advl	S – V – O(d)
	V – O(d)	Advl – V – Advl	S – V – Advl
		S – V – O(d)	O(d) – V – O(id)
		O(d) – V – Advl	
		Advl – V – O(id)	
		V – O(d)	
		O(d) – V – O(id)	

---

[2] https://elar.soas.ac.uk/Collection/MPI1083777 and Lingvodoc platform http://lingvodoc.ispras.ru/ (both accessed September 30, 2020).

Texts from 1868	Texts from 1886	Texts from 1946–1966	Texts from 2007–2010
	Attr [*O(d)] Advl – Attr	Advl – V – S	

There are a few examples of sentences without verbs; although the verb is reconstructed from the context, its exact position cannot be restored at the moment, so these examples need further study.

As follows from Table 2, the speech of the last speaker of the Lower Chulym dialect in the 21st century on the whole reflected the structures recorded in the middle and the end of the 19th century. At the same time, there were inventions that appeared only in the middle of the 20th century (e.g., direct word order – SVO). The subject was expressed more often in the last recorded texts than in the formerly documented ones. As noted above, the texts from the 19th century were a folklore story and a translation.[3] Whereas the texts from the 21st century were narratives, the text pattern in terms of the word order was often similar.

Due to space considerations, I will only list a few examples of sentence word order patterns from the Lower Chulym Turkic texts with limited analysis here.

Advl – S – O(id+d) – V

The adverbial modifier (of time) is fronted with the following subject, indirect and direct objects and verbal predicate:

(1) **Ӱш јылдың пажында Тар Пäг Моңул Канға албан аппар-тыр**

[Radloff 1868]

Ӱш јыл-дың паж-ын-да Тар Пäг Моңул Кан-ға албан аппар-тыр
Üš jïl-dïŋ paž-ïn-da Tar Päg Moŋul Kan-ya alban appar-tïr
3 year-GEN head-POSS.3SG-LOC Tar Päg Moŋul Kan-DAT tribute bring-PRS
Advl(t).[NUM+N+N] S.[N+N] O(ind).[N+N] O(d).[N] V.[V]
'At the beginning of every three years, Tar Päg pays tribute to Mongul-Kan.'

---

3 The inclusion of the Bible translation – a translated, not documented, text per se – into the current analysis may be questioned and criticized. However, there are very few Lower Chulym texts on the whole, and I use all the texts I have found or recorded in my analysis here. Thus, distinction between the results of documented and translated text analysis is to be left for further study.

O(d) – Advl – O(id) – V

The direct object is fronted with the adverbial modifier (of time) followed by the indirect object and verbal predicate:

(2) **Күннүг ажыбісті бугүн біске бергіл** [TEV 1886]
Кӱн-нӱг аж-ыбіс-ті бугӱн біс-ке бер-гіл
Kün-nüg až-ïbis-ti bugün bis-ke ber-gil
day-ADJ bread-POSS.1PL-ACC today we-DAT give-IMP
O(d).[ADJ+N] Advl(t).[ADV] O(ind).[PRO] V.[IMP]
'Give us this day our daily bread.'

S – V – O(d)

The subject is followed by the verbal predicate and the direct object (which is also defined by the final attribute):

(3) **Ол пилвӓ:н тӓренгини ол ораныӈ** [Dulzon 1952]
ол пил-вӓ:-н тӓренг-ин-и ол ора-ныӈ.
ol pil-vä:-n täreŋ-in-i ol ora-nïŋ
he know-NEG-PST depth-POSS.3SG-ACC that pit-GEN
S.[PRO] V.[V] O(d).[N] Attr.[PRO+N]
'He did not know the depth of that pit.'

The word order in the example above displays the attribute (*ol oranïŋ* 'of that pit') following the object it defines (*täreŋini* 'the depth') and may be understood as a calque from Russian, which in the original source is: *и он не узнал глубину этой ямы*, lit. 'and he not found_out the_depth of_this pit' (Dulzon 1952).

Advl – S – O(d) – V

The adverbial modifier (of time) precedes the subject followed by the direct object and the verbal predicate:

(4) **Qištinda ol meni kält'jadïyan** [Lemskaya 2007–2010]
qištin-da ol meni kält'-ja-dïyan
wintertime-LOC 3SG 1SG.ACC make.come-AUX:lie-PST
Advl(t).[N] S.[PRO] O(d).[PRO] V.[V]
'In winter, he would make me come'

2.3 Word order variation in Middle Chulym Turkic

In order to correlate with the data for analysis selected from the Lower Chulym Turkic material, I will take Tutal Middle Chulym Turkic texts of different time periods and make a random selection of 20 simple sentences within each text group; Melet Middle Chulym Turkic texts will not be considered in this article due to both size limitations and lack of diachronic material for comparison since the earliest recorded texts of the Melet subdialect go back to as early as 1971, whereas there were already published Tutal texts in 1886 (TEV) and 1909 (Malov).

The word order patterns of the Middle Chulym Turkic texts are summarized in Table 3:

Table 3: A comparison of Middle Chulym Turkic simple sentence word order types from a diachronic perspective.

Texts from 1886	Texts from 1909	Texts from 1970–1971	Texts from 2016
Advl – O(d) – V	Advl – O(d) – V	Advl – S – O(d) – V	Advl – S – O(d) – V
Advl – S – V	Advl – S – V	Advl – S – O(d) – Advl – V	Advl – S – V
O(d) – Advl – V	S – Advl – V	Advl – O(d) – V	S – Advl – O(d) – V
O(d) – Advl – O(id) – V	Advl – O(d) – Advl – V	Advl – S – V	S – O(id+d) – V
O(d) – (O(id) –) V	S – O(d) – V	O(d) – Advl – V	Advl – V
S – V	O(d) – V	S – Advl – V	
		S – O(id) – Advl – V	
		S – O(d/id) – V	
		O(d) – S – V	
		O(d+id) – V	
		S – V	
Advl – V – O(id)	Advl – O(d) – V – O(id)	Advl – S – V – O(id)	Advl – S – V – Advl
V – O(id)	Advl – V – Advl	Advl – V – O(d)	S – Advl – V – Advl
		S – V – O(id)	V – Advl
		V – O(d)	V – Advl – O(d)
			V
			Advl – V – S
			V – S
			Advl – S
O(id)			
Attr – O(id)			
Advl – O(id)			

Similar to the Lower Chulym cases, there are sentences without the expression of verbs, which need further study.

Table 3 shows that there are many more sentence types to be distinguished in Middle Chulym Turkic than in Lower Chulym, and the former attest a greater degree of variation in these types among the texts recorded in different time periods. However, there is a number of common features, as well, like quite a strong tendency of placing the adverbial modifier in the initial position. The expression of the subject is more common in the materials starting from the middle of the 20th century as is the reposition of the predicate (with various sentence types where the predicate immediately follows the subject).

Here are a few examples of sentence word order patterns from the Middle Chulym Turkic texts.

S – V

The subject is followed by the verbal predicate:

(5)   *Аны̄ң пашкарлыгыны̄ң учу болбос*   [TEV 1886]
       Аны̄ң пашкарлыг-ы-ны̄ң учу бол-бос
       Anïŋ paškarlïɣ-ï-nïŋ uči bol-bos
       DEM.GEN kingdom-POSS.3SG-GEN end be-NEG
       S.[PRO+N+N] V.[V]
       'Whose kingdom shall have no end.'

Here, the syntactic pattern of the translated text is not a calque of the original Russian one: *Его же Царству не будет конца* (lit.) 'His though Kingdom will not (have) an end', where the sentence does not have a grammatical subject, and the verbal predicate is surrounded by the direct and indirect objects.

Advl – S – V

The adverbial modifier (of time) is fronted and followed by the subject and the verbal predicate:

(6)   *Аның сонда тӱтӱн боп-парған*   [Malov 1909]
       Аның сон-да тӱтӱн бо-п-пар-ған
       anïŋ son-da tütün bo-p-par-ɣan
       DEM.GEN end-LOC smoke be-CVB-AUX:go-PST
       Advl(t).[PRO+N] S.[N] V.[V]
       'Then the smoke cleared.'

Advl – V – O(d)

The adverbial modifier (of time) is followed by the verbal predicate (with the omitted subject) and the direct object:

(7) **а:нынг сонда кӧрпаҮабыс моҮалак** [Abdrahmanov 1970]
 а:нынг сонда кӧр-па-Үа-быс моҮалак.
 a:nïŋ son-da kör-pa-ya-bïs moyalak
 DEM.GEN end-LOC see-AUX:go-PST-1PL bear
 Advl(t).[PRO+N] V.[v] O(d).[N]
 'After that we got to see a bear.'

S – Advl – O(d) – V

The subject occupies the initial position followed by the adverbial modifier (of manner) and the direct object with the verbal predicate at the final position:

(8) **adaj pazak anï köörübül** [Lemskaya 2016]
 adaj pazak anï köör-übül
 dog again 3SG.ACC see-PRS
 S.[N] Advl(m) O(d) V.[v]
 'The dog is looking at it again.'

3 Further study of Chulym Turkic word order variation

This serves as a compilation of Chulym Turkic sentence structures. Due to space limitations, my analysis dealt with designating only the sentence constituents and indicating their positions. Further study is necessary of the Chulym Turkic sentence from the viewpoint of information structure of its constituents and the semantic roles they play.

An illustration of such analysis for example (2) above may be represented as follows:

(9) Kün-nüg až-ïbis-ti bugün bis-ke ber-gil
 day-ADJ bread-POSS.1PL-ACC today we-DAT give-IMP
 | O(d) | ADV T | V
 'Give us this day our daily bread.'

This is not to be understood as a calque from the original Russian text (*хлеб наш насущный дай нам на сей день* [lit.] 'bread our daily give us for this day');

rather the Chulym Turkic word order here may be interpreted as where the verb attracts the indirect object and places it in the prior position to mark the Target.

4 Conclusions

The analysis of the simple sentence types in the Lower Chulym and (Tutal) Middle Chulym Turkic material from a diachronic perspective shows a large degree of variation in the syntactic structure among these dialects. However, there are quite a few common tendencies observed, like the fronting of the adverbial modifiers or repositioning of the predicate close to the subject. What is more, it may be inferred from the material that there was a strong tendency of a shift from the common Turkic SOV word order of the sentence towards the SVO order, possibly thus focusing on the Target[4] of the sentence. The reason for such a change might be a strong and long-term language contact with Russian (possible contacts with Paleosiberian languages must yet be studied separately); or such a change may result from other tendencies like the emphatic forwarding of the focus, or the addition of the formerly eliminated word (group) at the end of a clause, which on the whole may also be an internal characteristic of SOV languages (cf. Ponaryadov 2010). The information structure (the topic vs. focus of the sentence, etc.) may be indicated via phonetic features like stress, pitch, and intonation (cf. Tokmashev 2020), a topic yet to be studied for Chulym Turkic. The word order of the Chulym Turkic sentence must be further analyzed from the perspective of the actual division of the sentence, in view of its pragmatic content and, where possible, with the analysis of audio recordings as a priority.

Acknowledgments: I sincerely thank the Russian Science Foundation for the grant "Sozdaniye elektronnogo dialektologicheskogo atlasa tyurkskih yazykov Rossii" [Creating a Digital Dialect Atlas of Russia's Turkic Languages (mapping)] (Project 18-18-00501); this publication was prepared within the framework of this research project.

---

4 However, Targets in Chulym Turkic regularly appear before the verb in contrast to Turkic languages in Iran (see Bulut, this volume). For a definition of 'Target' see Asadpour (this volume).

Abbreviations

1, 2, 3	1st, 2nd, 3rd person
ACC	accusative case
ADJ	adjective
ADV	adverb
Advl	adverbial
Advl(m)	adverbial modifier of manner
Advl(t)	adverbial modifier of time
Attr	attributive
AUX	auxiliary verb
DAT	dative case
DEM	demonstrative form
GEN	genitive case
IMP	imperative
LOC	locative case
N	noun
NEG	negative
NUM	numeral
O	object
O(d)	direct object
O(id)	indirect object
PL	plural
POSS	possessive
PRO	pronoun
PRS	present tense
PST	past tense
S	subject
SG	singular
T	Target
V	verbal predicate
v	verb

References

Abdrahmanov, Makhmud A. 1970. Teksty chulymsko-tjurkskogo jazyka (sredne-chulymskij dialekt). [Texts of the Chulym Turkic Language (the Middle Chulym Dialect)]. *Jazyki i toponimija Sibiri* 2. 58–69.

All-Russian Census of 2010. National Composition and Language Proficiency, Citizenship. Vol. 4 Federal State Statistics Service. URL: https://www.gks.ru/free_doc/new_site/perepis 2010/croc/perepis_itogi1612.htm. Accessed 06.06.2020.

Anderson, Gregory D. S. & David Harrison K. 2006. Ös tili: Towards a comprehensive documentation of Middle and Upper Chulym dialects. *Turkic Languages* 10(1). 47–72.

Belikova, Olga B. 2001. Srednevekovaja istorija Zemli Pervomajskoj [The medieval history of the Pervomayskaya land]. In Ja. A. Jakovlev (ed.), *Zemlja pervomajskaja: sbornik nauchno-populjarnyh ocherkov*, 70–93. Tomsk: Izdatel'stvo Tomskogo universiteta.

Biryukovich, Rimma M. 1971. *Materialy po jazyku chulymskih tatar* [Materials on the language of the Chulym Tatars]. T. III. Teguldet. 658; T. VI. Pasechnoye. 361.

Boni, Roza A. 1973. *Materialy po jazyku chulymskih tatar* [Materials on the language of the Chulym Tatars]. T. VII. Tegul'det. 584.

Boyarshinova, Zoya Ya. 1950. Naselenie Tomskogo uezda v pervoj polovine XVII veka [The population of Tomsk County in the first half of the 17th century]. Trudy Tomskogo gosudarstvennogo universiteta. T. 112. Tomsk. 36–199.

Comrie, Bernard. 1989. *Language universals and linguistic typology. Syntax and morphology.* 2nd edition. Basil Blackwell.

Dulzon, Andrey P. 1952. Chulymskie tatary i ih jazyk [The Chulym Tatars and their language]. *Uchenye zapiski Tomskogo gosudarstvennogo pedagogicheskogo instituta.* T. IX. 76–211.

Dulzon, Andrey P. 1966. Chulymsko-tjurkskij jazyk [The Chulym Turkic language]. Jazyki narodov SSSR, T. 2, *Tjurkskie jazyki*, 446–466.

Dulzon, Andrey P. 1973. Dialekty i govory tjurkov Chulyma [The dialects and the idioms of the Turks of the Chulym]. *Sovetskaja Tjurkologija* 1973–2. 16–29.

Georgi, Johann G. 1780. *Russia: Or, a complete historical account of all the nations which compose that empire.* Vol. 2. London: printed for J. Nichols, T. Cadell, in the Strand; H. Payne, Pall-Mall; and N. Conant, Fleet-Street.

Kostrov, N. A. 1828. Zhenshchina u inorodtsev Tomskoy gubernii. 42. [The Woman among the Indigineous Dwellers in the Tomsk Province].

Lemskaya, Valeriya M. 2007–2010. *Lower Chulym Turkic field records.* Kaftanchikovo.

Lemskaya, Valeriya M. 2006–2018. *Middle Chulym Turkic field records*. Pasechnoye, Teguldet, Tomsk.

Lemskaya, Valeriya M. 2010. Chulymskie teksty [Chulym Turkic texts]. *Annotirovannye fol'klornye teksty obsko-enisejskogo jazykovogo areala*, 263–314. Tomsk: Veter.

Lemskaya, Valeriya M. 2012. Chulymsko-tjurkskij tekst [A Chulym Turkic text]. *Sbornik annotirovannyh fol'klornyh i bytovyh tekstov obsko-enisejskogo jazykovogo areala.* T. 2, 184–237. Tomsk: Agraf-Press.

Lemskaya, Valeriya M. 2013. Chulymsko-tjurkskij tekst "Krasivyj-korichnevyj" [A Chulym Turkic text "Beautiful Brown"]. *Sbornik annotirovannyh fol'klornyh i bytovyh tekstov obsko-enisejskogo jazykovogo areala.* T. 3, 295–345. Tomsk: Vajar.

Lemskaya, Valeriya M. 2015. Chulymsko-tjurkskie teksty [Chulym Turkic texts]. *Sbornik annotirovannyh fol'klornyh i bytovyh tekstov obsko-enisejskogo jazykovogo areala. Kollektivnaja monografija.* T. 4, 217–291. Tomsk: TML-Press; Vajar.

Lemskaya, Valeriya M. 2016. Chulym Turkic people and their dialects. In S. Eker & Ü. Ç. Şavk (eds.), *Endangered Turkic languages II A: Case studies*, Vol. 2, 93–115. International Turkic Academy. Hodja Akhmet Yassawi International Turkish-Kazakh University. Ankara Astana.

Lemskaya, Valeriya M. 2017. Chulymsko-tjurkskie teksty [Chulym Turkic texts]. *Sbornik annotirovannyh fol'klornyh i bytovyh tekstov obsko-enisejskogo jazykovogo areala*, 103–177. T. 5. Tomsk: TML-Press; Vajar.

Lemskaya, Valeriya M. 2020a. Chulymsko-tjurkskie teksty [Chulym Turkic texts]. *Sbornik annotirovannyh fol'klornyh i bytovyh tekstov obsko-enisejskogo jazykovogo areala.* T. 6, 233–276. Tomsk: Agraf-Press; Vajar.

Lemskaya, Valeriya M. 2020b. Porjadok slov v nizhnechulymskom dialekte chulymsko-tjurkskogo jazyka: diahronicheskij aspect [Word order in the Lower Chulym dialect of the Chulym-Turkic language: A diachronic aspect]. *Tomsk Journal of Linguistics and Anthropology.* 4(30). 41–59.

Tomskie Eparkhial'nye Vedomosti. 1886. Poezdka Preosvjashhennago Makarija, Episkopa Bijskago, k Chulymskim inorodcam Mariinskago okruga [The trip of Preconsecrated Makariy, the Bishop of Biysk, to the Chulym indigenous dwellers of the Mariinsky District]. *Tomskija eparhial'nyja vedomosti.* 8. 15–21. Available online at: http://sun.tsu.ru/mminfo/000349391/index.html (accessed September 9, 2020).

Malov, Sergey E. 1909. Otchet o komandirovke studenta Vostochnogo fakul'teta Sergeja Efimovicha Malova [A report on the trip of a student of the Oriental Department, Sergei Efimovich Malov]. *Izvestija Russkogo komiteta dlja izuchenija Srednej i Vostochnoj Azii v istoricheskom, arheologicheskom i antropologicheskom otnoshenijah,* Vol. 9, 35–46. Saint Petersburg.

Pomorska, Marzanna 2004. *Middle Chulym noun formation.* Kraków: Ksiegarnia Akademicka. 256.

Ponaryadov, Vadim V. 2010. *Porjadok slov v permskih jazykah v sravnitel'no-tipologicheskom osveshhenii (prostoe predlozhenie)* [The word order in Perm languages in a comparative typological coverage (the simple sentence)]. Syktyvkar. 120.

Radloff, Wilhelm. 1868. *Taska Maatyr. Die Sprachen der Türkischen Stämme Süd-Sibiriens und der Dsungarischen Steppe.* I Abtheilung. Proben der Volkslitteratur. II. Theil. Saint Petersburg. 689–705.

Radloff, Wilhelm. 1989. *Iz Sibiri: Stranitsy dnevnika* [From Siberia: Pages from a diary]. Moscow. 749.

Strahlenberg, Philipp J. von. 1730. *Das nord- und ostliche Theil von Europa und Asia...* Stockholm. 438.

Schönig, Claus. 1997. A new attempt to classify the Turkic languages (1) [Text] / C. Schönig // *Turkic Languages* 1. 117–133.

Tokmashev, Denis M. 2015. *Middle Chulym Turkic field records.* Pasechnoye.

Tokmashev, Denis M. 2020. Informacionnaja struktura prostogo predlozhenija v teleutskom jazyke: k postanovke problem [Information structure of a simple sentence in the Teleut language: To the statement of the problem]. *Bulletin of Kemerovo State University.* 22(1). 268–277.

Zemlja Pervomajskaja. 2001. sbornik nauchno-populjarnyh ocherkov / ed. by Ja. A. Jakovlev. Tomsk: *Izdatel'stvo Tomskogo universiteta.* 548. [The Pervomayskaya Land: a Collection of Popular Science Essays].

Daniel Birnstiel
Copulas and Target phrase positioning in the Arabic dialects of Kurdistan

Preliminary remarks on two information-structure related features

Abstract: This article deals with the phenomenon of copulas and the word order variation of Target phrases in certain varieties of Arabic spoken in Kurdistan. After discussing Mesopotamian Arabic and its subgroups from a general perspective, the various types of copulas are introduced and exemplified together with an analysis of some of the functions that can be observed. The next section analyses the placement of selected Target phrases and discusses the differences in meaning resulting from this word order variation. A brief reflection on linguistic contact as cause for the development of these characteristics of Mesopotamian Arabic precedes the preliminary conclusions.

Keywords: incorporation; information structure; presentatives; Target phrases

1 Introduction

This contribution addresses both the question of copulas and the placement of Target expressions in the Arabic dialects spoken in Kurdistan.[1] I precede the discussion of these features with some remarks concerning the Arabic dialects of Kurdistan in general (Section 2). These include comments on the established terminology, information concerning the distribution of dialect groups, subgroupings, and the coexistence of communal dialects.

Following a general description of copulas as a grammatical category in the Arabic dialects of Kurdistan (Section 3.1), I offer some observations regarding

---

[1] For definitions of the terms "copula" and "Target" as used in this paper refer to the respective sections below.

Daniel Birnstiel: Institute for the Study of Islamic Culture and Religion, Johann Wolfgang Goethe-University, Post box 11 19 32, 60054 Frankfurt am Main, Germany,
E-Mail: birnstiel@em.uni-frankfurt.de

several of the attested functions in some dialects and discuss the use of copulas with a focus on information structure (Section 3.2).

The subsequent section concerns the placement of certain Target phrases denoting recipients of giving and taking (Section 4.1) as well as addressees of verbs of speaking (Section 4.2). The positions taken by these Targets are exemplified and the function of word order variation is analyzed.

A short discussion of the impact of language contact on the development of these features (Section 5) precedes the general conclusions (Section 6).

2 The Arabic dialects of Kurdistan and adjacent regions

This article deals primarily with varieties of Arabic spoken in the Kurdistan linguistic area, i.e., the geographic region at the convergence of the modern nation states of Armenia, Azerbaijan, Georgia, Iran, Iraq, Syria, and Turkey. The dialects of Arabic spoken in this linguistic contact zone have been in close contact with several languages, especially the different varieties of Kurdish and other Iranian languages, different North-Eastern Neo-Aramaic languages, as well as Turkish and other Turkic languages such as Azeri or Turkoman (Procházka 2020). In earlier times, Arabic was also in close contact with languages to the south of Kurdistan, especially with different varieties of Eastern Aramaic.

2.1 Arabic dialect classification according to geography

Arabic dialects can be classified according to several different criteria. The most prevalent of these is classification by geographical factors, i.e., according to regions or modern nation states. Although dialect groupings by geographic location provide clear and convenient labels, the various dialects grouped under a given label do not necessarily form a dialect bundle connected by shared isoglosses or a common origin (Versteegh 1997: 145; Palva 2006–2009: 604). Geography-based terms that refer to or include varieties of Arabic spoken in Kurdistan are Anatolian Arabic, Iraqi Arabic and Mesopotamian Arabic.

Mesopotamian Arabic is a cover term for all varieties of Arabic attested in Mesopotamia, i.e., the region covered by the Euphrates and Tigris rivers with their tributaries. This comprises all of Iraq, south-eastern Turkey (Anatolia), north-eastern Syria and the region of Khuzestan in south-western Iran. It is the established term for one of the five major Arabic dialect groups.[2] The Arabic dialects of Mesopotamia can be divided into two major dialect subgroups, namely *qəltu* and *gəlet* dialects (see Section 2.2 below). Nevertheless, they can be said to form a unit on more than purely geographic grounds. Blanc (1964: 6f.) lists eleven isoglosses shared by most Mesopotamian dialects irrespective of which of these two groups they belong to.[3]

Two other geography-based terms that refer to or include Arabic dialects of Kurdistan are Anatolian Arabic and Iraqi Arabic. However, these are only partially suited when discussing the possibility of language contact-induced features in the geographic region described above. The adjective "Iraqi" refers to a modern nation state, while "Anatolian" refers to several geographic regions of modern Turkey.

Admittedly, the Arabic dialects of Anatolia indeed form a cohesive group of dialects, albeit one that is closely related to varieties spoken to the south in northern Iraq, all of which belong to the *qəltu* subgroup of the two Mesopotamian dialect groups mentioned above.

The Arabic dialects of Iraq, on the other hand, belong to both of these subgroups. However, they are spoken in all parts of Iraq, i.e., also outside Iraqi Kurdistan, and indeed beyond the borders of Iraq. Consequently, the term "Iraqi Arabic" does not correspond to any clear linguistic isoglosses that would set Iraqi Arabic apart from Anatolian Arabic or establish it as a cohesive, independent group within Mesopotamian Arabic.[4]

---

[2] The other four are (1) the dialects of the Arabian Peninsula, (2) the dialects of the Levant, (3) the dialects of Egypt and Sudan, and (4) the dialects of North Africa (Behnstedt et al. 1980: 23–38; Versteegh 1997: 145, 2014: 189; Talay 2011: 909; Watson 2011, especially 853–856).

[3] These isoglosses include (1) the emergence of [p] as a phoneme, (2) the preservation of the Old Arabic interdentals [θ], [ð] and [ðˤ] in most cases, (3) the realization of Old Arabic [q] as a uvular stop ([q] or [ɢ]), opposed to its frequent realization as a glottal stop ([ʔ]) in sedentary Levantine dialects, (4) the preservation of final *-n* in the 2nd/3rd person (masculine) plural and 2nd person feminine singular endings of the Old Arabic imperfect (= present/future), (5) the emergence of an indefinite article, and (6) the emergence of the particles of existence *āku* 'there is' and *māku* 'there isn't'.

[4] Despite the shortcomings of these terms, the *Encyclopedia of Arabic Language and Linguistics* has entries on "Anatolian Arabic", "Iraq", and "Khuzestan Arabic", but not "Mesopotamian Arabic" (Ingham 2006–2009; Jastrow 2006–2009a; 2006–2009b).

2.2 *qəltu* and *gəlet* dialects

Another important factor is the distinction between sedentary and Bedouin dialects (Palva 2006–2009: 605f.). Sedentary dialects are further divided into urban and rural dialects; moreover, when Bedouin tribes settle, their dialects often undergo "sedentarization" as a result of increased contact with speakers of sedentary dialects, while many sedentary, especially urban dialects, undergo "Bedouinization" due to the influx of speakers of Bedouin dialects (Palva 2006–2009: 605f., 611f.).

This distinction between sedentary and Bedouin dialects lies at the basis of the bifurcation of Mesopotamian dialects into *qəltu* and *gəlet* (or *gilit*) dialects. The terminology utilizes the respective dialect reflexes of the Old Arabic verb form *qultu* 'I said' to illustrate two major isoglosses distinguishing these two dialect groups: (1) the realization of the Old Arabic voiceless uvular stop [q] as a voiceless uvular stop [q] in the *qəltu* dialects but as a voiced uvular stop [ɢ] in the *gəlet* dialects, and (2) the realization of the inflection suffix of the 1st person singular past[5] as *-tu* in the *qəltu* dialects and as *-ət* in the *gəlet* dialects (Blanc 1964: 5f.; Jastrow 2006–2009b: 414; Talay 2011: 909f.). Dialects of both groups are spoken beyond the borders of Iraq and Anatolia.

The *qeltu* dialects represent a relatively old linguistic layer. All Christian and Jewish dialects of Mesopotamia as well as all dialects spoken by sedentary Muslim populations north of the line between Samarra and Falluja belong to this dialect group; their territory stretches beyond the borders of Iraq into south-eastern Anatolia and north-eastern Syria. They exhibit some affinities with Uzbekistan Arabic, the speakers of which originate from Iraq (Jastrow 2006–2009b: 414; Talay 2011: 910).

The *gəlet* dialects on the other hand represent a relatively recent linguistic layer in the region. They are spoken by nomadic, sedentarized, and Bedouinized (Muslim) populations throughout Iraq as well as sedentary Muslim populations south of the Samarra-Falluja line. Dialects belonging to this group are also attested in Khuzestan (Iran), Kuwait, and north-eastern Syria. They show affinities with the Bedouin dialects of the Syrian desert and the northern Arabian Peninsula (Jastrow 2006–2009b: 414; Talay 2011: 910).

The *gəlet* dialects are generally thought to have migrated into Mesopotamia in the aftermath of the Mongol invasion (1258–1400), reaching a peak during the Ottoman rule (1533–1918), with the influx of (former) Bedouin populations into

---

5 I.e., the "perfect" in Semitist or Classical Arabic terminology.

urban centers causing dialect shifts among the sedentary Muslim populations of the south (Talay 2011: 910).

Following Jastrow (2006–2009b; 2006–2009a; see also Talay 2011: 910f.), the *qəltu* dialects can be divided into four major groups (Anatolian, Tigris, Euphrates, and Iraqi Kurdistan),[6] each of which comprises several subgroups, while the *gələt* dialects may be classified into three groups (northern Mesopotamian, central Iraqi, and southern Iraqi and Khuzestan) according to Talay (2011: 911f.). The table in Section 2.4 lists these groups together with additional information concerning denominational affiliation, geographic location, etc.

2.3 Communal dialects

Dialect classification according to religious denomination refers to the phenomenon of coexisting communal dialects (Palva 2006–2009: 610). This means that populations from the same place may speak different dialects when belonging to different religious communities (Jastrow 2006–2009b: 414; Talay 2011: 910; Watson 2011: 854). The textbook example for this phenomenon is the situation in Baghdad described by Blanc (1964: 8–10), where two *qəltu* dialects, a Christian and a Jewish one, are attested alongside a Muslim *gələt* dialect.

This phenomenon is also attested elsewhere and may extend to different denominations within the same religious community, e.g., Karaite and Rabbinic Jews, Shia and Sunni Muslims, or different Christian denominations. Thus, e.g., Shia and Sunni communal dialects can be distinguished in Bahrain (Holes 2006–2009: 241) while Karaite and Rabbinical Jewish dialects are attested in Iraq (Khan 1997).

The existence of community-specific dialects in the same locality attests usually to a high degree of historic mobility. That is to say, the coexistence of different Christian, Jewish and/or Muslim Arabic dialects in the same locality is often due to the fact that the members of one of the communities had migrated from elsewhere.

For example, the Sunni Arabic dialects of Bahrain are spoken by the descendants of tribes that migrated to the area from the Najd region of northeastern Arabia (Holes 2006–2009: 241). Similarly, the situation in Baghdad was

---

6 Jastrow actually uses the term "Kurdistan group" for the fourth group of *qəltu* dialects. Given the fact that Kurdistan as defined above extends well into Anatolia, I have amended Jastrow's terminology to "Iraqi Kurdistan" with the subgroups "northern Iraqi Kurdistan" and "southern Iraqi Kurdistan" respectively.

influenced by the migration of Muslim speakers of a *gələt* dialect who in turn caused the sedentary Muslim population of the city to switch dialects eventually (Palva 2006–2009: 612; Versteegh 2014: 202). In other cases, members of the minority group may have migrated to avoid religious persecution, taking their own language with them. Community-specific language forms are also known from languages other than Arabic, e.g., Christian and Jewish (Neo-)Aramaic, Jewish and Zoroastrian Dari, Jewish Berber, Ladino and Yiddish.

2.4 Arabic in Mesopotamia and Kurdistan – working definition and data

For the purpose of this article, the Arabic dialects of Kurdistan are defined as a subgroup of Mesopotamian Arabic, irrespective of whether they are *qəltu* or *gələt* dialects and regardless of the urban, Bedouin, or religious origin of their speakers. Accordingly, Anatolian Arabic is one subgroup within Kurdistan Arabic along additional subgroups spoken in Iraq.[7]

Most of the data we possess for Mesopotamian Arabic (and thus for Kurdistan Arabic dialects) comes from Christian and Jewish *qəltu* dialects (Talay 2011: 910f.). This creates certain difficulties. On the one hand, the various political upheavals since the end of the Ottoman empire have caused large parts of the religious minority groups to migrate to the West as well as to Israel; this has either resulted in these dialects being severely threatened by language death or even having already become extinct (Talay 2011: 910), making it difficult if not impossible to complete the picture regarding certain attested linguistic features. On the other hand, the current situation in large parts of the area does not permit carrying out the fieldwork necessary for mapping the remaining minority varieties or the still largely undocumented Muslim *gələt* (as well as *qəltu*) dialects. All examples adduced in this article are taken from Jastrow's seminal survey (Jastrow 1978; 1981).[8]

---

[7] For a general overview regarding the state of research concerning the Mesopotamian dialects in general as well as the individual dialects surveyed, see the following with the literature referred to there: Ingham (2006–2009), Jastrow (2006–2009a; 2006–2009b), Palva (2006–2009).

[8] I have generally kept to Jastrow's transliteration, which is largely based on the convention of the *Deutsche Morgenländische Gesellschaft* (DMG). I have, however, decided to replace the Arabic character ع employed by Jastrow to represent the Arabic pharyngeal /ʕ/ with the corresponding DMG character ʻ. Also, I follow the Leipzig glossing rules regarding the marking of clitic boundaries with the equals sign but affix boundaries with hyphen. I treat object pronouns and the definite article as affixes, but phonetically shortened forms of prepositions as clitics.

The following table lists the various Arabic dialect groups of Mesopotamia according to their subgroupings and location:[9]

Table 1: Arabic dialects of Mesopotamia.

Dialect type	Group	Subgroup	Dialects/Localities	Denomination	Modern state
qəltu	Anatolian	Mardin	Mardin town, Mardin villages, Plain of Mardin	Christians, Muslims	Turkey
			Kōsa, Mḥallami	Muslims	
			Āzəx	Christians	
			Nusaybin, Cizre	Jews	
		Siirt	Siirt town	Christians, Muslims	
			Siirt villages	Muslims	
		Diyarbakır	Diyarbakır town	Christians, Jews	
			Diyarbakır villages	Christians	
			Siverek, Çermik, Urfa	Jews	
		Kozluk-Sason-Muş	Kozluk, Sason, Muş, Daragözü	Christians, Muslims	
	Tigris	Mosul and surroundings	Mossul, Bəḥzānī, Baʿšīqa, ʿAyn Səfne	Christians, Jews, Muslims, Yezidis	
		Tikrit and surroundings		Muslims	
		Baghdad and southern Iraq		Christians, Jews	
	Euphrates	Khawētna-tribe	əl-Khatūnīye, əl-Hōl and surrounding villages	Muslims?	Syria, Iraq, Turkey

---

9 This table is based on the data given in Jastrow (2006–2009a; 2006–2009b), Talay (2011) and Behnstedt (1992). The lack of specific dialects and localities for certain subgroups reflects the lack of data for these dialects. Concerning the Arabic dialects of Khuzestan, see the data given by Ingham (2006–2009) and Leitner (2020).

Dialect type	Group	Subgroup	Dialects/Localities	Denomination	Modern state
		Dēr iz-Zōr[10]		Muslims?	Syria
		'Āna- 'Albu Kmāl	'Ana	Jews, Muslims	Iraq
			'Albu Kmāl	Muslims?	Syria
		Hīt		Jews, Muslims	Iraq
	Iraqi Kurdistan	Northern Iraqi Kurdistan	Səndōṛ, ʿAqra, Arbil, Šōš	Jews	Iraq
		Southern Iraqi Kurdistan	Kirkuk, Tūz Khurmātu, Khānaqīn	Jews	Iraq
qəltu	Northern Mesopotamian	Syrian Šāwī	along Euphrates including Urfa, Raqqa	Muslims	Syria, Iraq, Turkey
		Rural dialects of Northern/Central Iraq			Iraq
	Central Iraqi	Muslim Baghdadi	Baghdad		
		Sunni area around Baghdad			
	Southern Iraqi/ Khuzestani	Urban dialects			Iraq, Iran
		Rural dialects			
		Marshland dialects			

3 The copula in Arabic dialects of Mesopotamia: introductory remarks

Many of the *qəltu* dialects of Mesopotamia have developed a copula. This is particularly true of the Anatolian dialects as well as other varieties of Arabic spoken in Kurdistan, e.g., Bəḥzānī, and even beyond; Christian Baghdadi, Uz-

---

10 Jastrow (Jastrow 1978; 1981) and others also write Dēr izZōr. In this article, however, the definite article is separated by hyphen in keeping with usage in other place names as well as the language examples.

stan Arabic and Maronite Cypriot Arabic also feature a copula (Blanc 1964: 124f.; Jastrow 2006–2009a: 91f., 2006–2009b: 419; Zemer 2008; Walter 2020: 169f.). Both Uzbekistan Arabic and Maronite Cypriot Arabic are in fact closely related to the Mesopotamian *qəltu* dialects and belong to the *qəltu* dialect continuum despite their status as isolated language islands (Jastrow 2006–2009b: 414; Palva 2006–2009: 612; Talay 2011: 909).

In this article, copula refers to a grammaticalized clitic element (or elements) related to or derived from (personal as well as other) pronouns and attached to the predicate in nominal clauses (Jastrow 1978: 131–141, especially 131).[11] Whereas Classical Arabic as well as different modern dialects do not use an explicit marker of predication, e.g., in the case of present tense nominal clauses, many of the Arabic varieties spoken in Kurdistan (and other parts of Mesopotamia) possess one or more sets of copulas. Copula here means an overt expression of the predicative relationship between subject and predicate.

I do not subscribe to the view that postulates the existence of a zero-copula in nominal clauses without an overt copula in Arabic[12] or Semitic in general. Rather, non-verbal predication in Semitic does not need to be expressed through a copula but may instead rely on other means, such as the lack of determination on the predicate (together with gender-number agreement in the case of adjectives), word order or simple juxtaposition. Similarly, I do not regard personal pronouns in nominal clauses with definite predicates as copulas, but rather as occupying the position of the subject, while the "logical" subject is a topic in extraposition.[13] However, an obligatory personal pronoun occupying the actual subject slot must at one time have presented an intermediate step in the development of copulas of the type found in the Arabic dialects of Kurdistan. Parallel developments are also known from other Semitic languages in the region, notably Syriac[14] and various varieties of Neo-Aramaic.[15] What characterizes the copulas in the dialects under discussion is their obligatory occurrence,

---

[11] Similar copulas are known from North-Eastern Neo-Aramaic (Khan 2006: 2012).
[12] This is for instance the approach taken by Camilleri and Sadler (2019).
[13] In Classical Arabic grammar theory these are treated as compound clauses, in which the extraposed subject is followed by a nominal clause. See Fischer (2001: 191–193). An example would be: *ulā'ika humu l-kāfirūn* (those they the-unbelievers) 'Those (people), they are the unbelievers'.
[14] See Goldenberg (1998).
[15] See Khan 2006: 2012).

their marking of the predicate and their agreement with the actual pronominal or nominal subject.[16]

3.1 Types of copulas

Based on the description in Jastrow (1978: 131–141),[17] the copulas extant in the Arabic dialects of Kurdistan can be classified according to the following parameters:
a) the position of the copula: with the exception of the Siirt dialects of Anatolia (Jastrow 1978: 132f.), the (positive) copula is enclitic to the predicate;
b) the grammatical person/s for which a copula/s exist/s: in most instances, the copula exists across the entire paradigm, i.e., different forms are available for all grammatical persons;
c) the form of the copula: allomorphs depending on the phonetic environment (post-vocalic or post-consonantal) may exist as well as shortened variants restricted to certain lexemes occurring as predicates;
d) number and types of copula sets: copulas may be positive, negative, relative or demonstrative (see the following sections).

3.1.1 The positive copula

The type of copula attested most is the general positive type that attaches to the predicate in positive nominal clauses; it has developed from independent personal pronouns in unstressed positions (Jastrow 1978: 131–136).

In Classical and Modern Standard Arabic as well as many modern dialects, present tense nominal clauses are formed without the need for a copula by mere juxtaposition of subject and predicate. In the Mesopotamian dialects that possess copulas, however, their usage becomes obligatory. Compare the following

---

[16] On this notion of copula in Arabic and other Semitic languages, see Goldenberg (2013: 153–158).
[17] Apart from Jastrow's publications covering the results of his fieldwork, descriptions of Mesopotamian Arabic usually do not present any information regarding these copulas beyond some generalizing description of the phenomenon.

examples from the dialect of Dēr iz-Zōr (Syria), which does not possess a positive copula, and the Mardin (Mḥallamī) dialect of Kənderib (Soğütlü, Turkey):[18]

(1) Dēr iz-Zōr [Jastrow 1978: 131)
    hal-walad   Ø   zēn
    this[19]-child:MSG   Ø   good:MSG
    'This child **is** good.'

(2) Kənderib [Jastrow 1978: 131]
    hal-walad   malīḥ=**we**
    this-child:MSG   good:MSG=**COP.3MSG**
    'This child **is** good.'

This distinction is even present where the subject is presented by an independent personal pronoun, e.g., in the following two examples:

(3) Dēr iz-Zōr [Jastrow 1978: 131]
    ana   Ø   zēn
    I   Ø   good:MSG
    'I **am** good/well.'

(4) Kənderib [Jastrow 1978: 131]
    ana   malīḥ=**ana**
    I   good:MSG=**COP.1SG**
    'I **am** good/well.'

The Siirt dialects, however, possess a copula which precedes the predicate.[20] This copula is only found with nominal subjects, in which case its use is obligatory; it does not occur with pronominal subjects (Jastrow 1978: 132):[21]

---

18 In the following examples, zero (Ø) denotes the absence of a copula not a zero-copula. See my remarks towards the end of the introductory section on copulas (Section 3).
19 This is the so-called demonstrative article, in which a demonstrative element *hā- has been fused with the following article; according to Jastrow, the demonstrative article has the same function as the demonstrative pronoun, with the latter supposedly carrying more force, although the choice may ultimately depend on the respective prosodic circumstances (Jastrow 1978: 110f.). Apparently, the issue awaits further clarification.
20 In his transliteration, Jastrow (1981: XIII) follows the tradition of Arabic orthography that writes prepositions and conjunctions that consist of more than two letters as independent words. Consequently, he treats prepositions that consist of more than a consonant followed by a short vowel as independent prosodic words, even though they usually do not carry an independent stress. Since stress is generally unmarked in the texts edited by him, it is difficult to decide if the copula in Siirt bears a stress or is possibly (pro)clitic to the following predicate.

(5) Siirt [Jastrow 1978: 132]
  āvī   l-bənt   **īye**   malīḥa
  this:FSG the-girl **COP.3FSG** good:FSG
  'This girl **is** good.'

(6) Siirt [Jastrow 1978: 132]
  īye   ∅   naxwašše
  she   ∅   ill:FSG
  'She **is** ill.'

These restrictions, i.e., the necessary occurrence of the copula in sentences with nominal subjects and its necessary absence from sentences with pronominal subjects, permit the distinction of copula and independent personal pronouns even though both have identical forms. Hence, only 3rd person copulas occur in this type of sentence. This indicates that the development of copulas in the Siirt dialects has not yet been completed, since there are no distinct copula forms that occur with explicit pronominal subjects.

Copula forms of the 1st and 2nd persons, however, are found in interrogative clauses featuring both yes-no questions and *wh*-questions. Here, the copulas (including the 3rd person) are enclitic and follow the predicate; the same is true in positive clauses where the predicate itself is an independent personal pronoun (Jastrow 1978: 132):[22]

(7) Siirt [Jastrow 1978: 132]
  əxt-ok   malīḥa=**ye**?
  Sister-your:MSG good:FSG=**COP.3FSG**
  '**Is** your sister fine?'

---

21 Elsewhere, however, Jastrow exemplifies his observation that the copula precedes the predicate in the Siirt dialects as opposed to the other Anatolian dialects with the supposedly minimal pair *ūwe malī* (COP.3MSG good:MSG) (Siirt) with *malēḥ-we* (good:MSG-COP.3MSG) (Mardin), both meaning "he is good", which seems to indicate that both examples feature a copula (Jastrow 1981: 218).

22 In the latter case, shortened variants of the 3rd and 1st person singular copulas are used, i.e. *-we* (3MSG), *-ye* (3FSG), *-na* (1SG) (Jastrow 1978: 132). I am somewhat confused by Jastrow's statement that the form of the copula is identical to the of the personal pronoun, which means the form of the 1st person should be *anā*, while pointing out that following a personal pronoun as predicate, the 1st person copula is shortened to *-na*. Presumably, the full copula form *ana* occurs in interrogative clauses. However, I was unable to find any examples of a 1st person copula in the Siirt corpus available to me.

(8) Siirt [Jastrow 1978: 132]
 mən=**ənten?**
 who=COP.2PL
 'Who **are you**?'

(9) Siirt [Jastrow 1978: 132]
 anā=**na**
 I=COP.1SG
 '**It's** me.'

3.1.2 Negative copulas

In certain dialects, however, additional copula sets exist. Many dialects also possess a negative copula in which a negating particle has been fused with an enclitic positive copula (Jastrow 1978: 136–138):

(10) Kəndērib [Jastrow 1978: 137]
 fə=l-bayt=**we**
 in=the-house=COP.3MSG
 '**He is** in the house.'

(11) Kəndērib [Jastrow 1978: 137]
 mawwe fə=l-bayt
 NEG.COP.3MSG in=the-house
 '**He is not** in the house.'

These negative copulas, e.g., *mawwe* in the preceding example, cannot be simply analyzed as a negating particle followed by the (synchronically) enclitic copula, e.g., -*we*. Rather, they are the results of a fusion between a negating particle followed by an enclitic form of the independent personal pronoun, where both elements subsequently underwent phonological changes, i.e., *mawwe* < *mā-huwwa*, etc.

Interestingly, the existence of a negative copula is not dependent on the existence of a positive copula; the dialect of Dēr iz-Zōr, e.g., possesses a complete negative copula paradigm despite not having any positive (or other) copula at all (Jastrow 1978: 138). Moreover, some dialects such as the dialect of Āzəx possess only forms for the 3rd person while other dialects (e.g., Mardin city, Arbəl, and certain dialects of the Kōsa and Mḥallami subgroups) do not possess a negative copula at all (Jastrow 1978: 138).

3.1.3 Relative copulas

Some Anatolian dialects, moreover, also feature a relative copula in which the relative clause marker has blended with the enclitic positive copula representing the subject of a nominal relative clause; this type of copula exists only for the 3rd person (Jastrow 1978: 138) and is restricted to nominal, i.e., verbless, relative clauses in which the subject of the clause refers to the head noun. The following examples from Qarṭmīn (Yayvantepe) and Kənderib (Söğütlü), a Mḥallami sub-dialect, both of which belong to the Mardin group of Anatolian dialects (Jastrow 1978: 11f., 15) illustrates this type of copula:[23]

(12) Qarṭmīn [Jastrow 1978: 138]
yazīdīyət **lanne** fə=s-sūri
Yazidis REL.COP.3PL in=the-Syria
'The Yazidis **who are** in Syria.'

(13) Kənderib [Jastrow 1981: 58]
ḥawk səfnēyāt əl-lawx **lənne** bayn
those wedges:PL the-other:PL REL.COP.3PL between
əs-səfnētayn yərxawn
the-wedges:DU loosen:PRS.3PL
'Those other wedges **which are** between the two (before-mentioned) wedges come loose.'

In these examples, the copula, which is the result of fusing two elements, expresses both the relativizer and the pronominal subject of a nominal clause. This distinguishes these dialects from Classical and Qur'anic Arabic, where the relative pronoun is simply followed by a prepositional phrase or an adverb and the pronominal subject of the verbless clause remains unexpressed, if it refers to the head noun:

(14) Qur'anic Arabic [Qur'an 22:46][24]
l-qulūbu llatī Ø fī ṣ-ṣudūri
the-hearts:MPL REL.FSG Ø in the-chests
'the hearts that[25] **(are)** in the chests'

---

23 The predicate in most cases if not all seems to consist of a prepositional phrase or an adverb.
24 Here, zero (Ø) denotes the absence of a subject, not a zero-subject.
25 Note that according to the agreement rules of Classical Arabic, attributive adjectives and predicates to inanimate nouns are usually feminine singular.

It is tempting to analyze the relative pronoun as fulfilling a double function in this case, namely as expressing the relativizer as well as the pronominal subject of the nominal clause. In fact, however, the relative pronoun in these cases is the (pronominal) head of an attributive expression consisting of a head followed by a prepositional phrase as "genitive" attribute, with the whole expression standing in apposition to a preceding explicit noun (see Goldenberg 2013: 226–262, especially 258–262, and the literature quoted there). Hence, strictly speaking, *llatī fī ṣudūr* does not really constitute a (relative) clause at all, but rather a type of "adjectivized" prepositional phrase.

In Classical Arabic, the subject of the relative clause is however more commonly expressed by an independent personal pronoun even where it refers to the head noun (Reckendorf 1921: 426–430, especially 428f.). This means that the relative clause contains a resumptive pronoun and the relative pronoun is followed by an actual nominal clause:

(15) Classical Arabic [Reckendorf 1921: 426]
r-raǧulāni   lladāni   **humā**   ʿinda-ka
the-men:DU   REL.MDU   **they:DU**   at-your:2MSG
'the two men that **are** with you.'

The two types of constructions in the preceding two examples are attested in Classical (and Qurʾanic) Arabic and non-Anatolian varieties of Arabic. They differ from the above-mentioned Anatolian type which employs a relative copula denoting both the relativizer and the subject of the verbless relative clause.

Anatolian Arabic resembles example (14) in so far as a single relativizing element is involved. It differs, however, since the Anatolian relative copula derives from a fusion of relativizer and enclitic copula. In (15) on the other hand, there are two elements, a relative marker indicating the clause's status as relative clause and an independent personal pronoun presenting the subject. It differs from Anatolian Arabic with regard to each grammatical role being presented by a separate morphological element, but it presents a state similar to the one from which the Anatolian relative copula ultimately derives via fusion.

3.1.4 Demonstrative copulas and presentatives

Lastly, many of the Kurdistan Arabic dialects as well as Bəḥzāni also possess a demonstrative copula which often has some form of predicative function;[26] in

---

[26] Similar copulas with a parallel origin exist in North-Eastern Neo-Aramaic dialects, where there are referred to as deictic copulas (Khan 2012: 111–114).

some Anatolian varieties even two or three sets of demonstrative copulas coexist (Jastrow 1978: 139ff.). The simple form consists of a prefixed demonstrative element *k-* that has been attached to the enclitic copula/former independent personal pronoun; in some dialects, e.g., Āzəx, it is restricted to the 3rd person, though most dialects attest the entire paradigm. This copula is said to have a presentative ("look, he is...") and/or actualizing ("he is right now") function:

(16) Qarṭmīn [Jastrow 1978: 140]
 ana **kana** xalf əd-dawāb
 I DEM.COP.1SG behind the-pack animal
 '**I am (right now)** behind the pack animal.'

In some dialects (e.g., Kəndērib, Qarṭmīn), a demonstrative element *-hā* ('here') is suffixed once or twice to the 3rd person forms of this copula; the forms with single suffixed *-hā* may be enlarged by the particle *-ne* which ultimately derives from an old adverb meaning 'here' (*-ne* < **-nē* < **-nā* < **-hunā*) (Jastrow 1978: 101, 140).[27] The forms with single suffixed *-hā* function as presentatives of near and general deixis whereas the forms with double suffixed *-hā* function as presentatives of far deixis; they always precede the item presented:

(17) Qarṭmīn [Jastrow 1978: 140]
 kwā́ zlām mayyət
 PROX.PRES.3MSG person dead:MSG
 '**There is** a dead person.'

(18) Qarṭmīn [Jastrow 1978: 141]
 kēhā́ maṛa
 DIST.PRES.3FSG woman
 '**Over there** is a woman.'

The 3MSG form of the demonstrative copula with single suffixed *-hā* is attested in almost all Anatolian dialects as an indeclinable presentative particle introducing a sentence, even in dialects that generally do not possess this type of copula:[28]

---

27 E.g., the forms attested in Qarṭmīn: 3MSG *kū-hā(-nā)* > *kwā́(ne)*, *kū-hā-hā* > *kōhā́*; 3FSG *kī-hā(-nā)* > *kyā(ne)*, *kī-hā-hā* > *kēhā́*; 3PL *kən-hā(-nā)* > *kənā́(n/ne)*; there seems to be no doubly suffixed form for the 3PL in this dialect.
28 Somewhat similar to Jastrow (1978) and Khan (2012), I treat the forms that are attested for all persons and are not restricted to sentence initial position as demonstrative copulas, albeit with some presentative force. Presentative elements occurring in initial position and inflecting only for the 3rd person or else having become a petrified particle are treated as presentatives.

(19) Mardin [Jastrow 1978: 141]
kwā ǧaw əl-ḥarāmīye
PRES.PART come:PST.3PL the-thieves
'**Look**, the thieves have come!'

Both this presentative particle as well as the demonstrative copula with additional suffixed demonstrative element(s) discussed in examples (17) and (18) appear in front of the element they present or the sentence they present. The simple demonstrative copula, on the other hand, can occur in front of the predicate even in cases where there is an overt nominal subject:

(20) Qarṭmīn [Jastrow 1978: 140]
əṭ-ṭōrči **kū** qəddām əl-baḥar
the-fisher DEM.COP.3MSG before the-sea
'The fisher **is (right now)** on the seashore.'

(21) Qarṭmīn [Jastrow 1978: 140]
kwā́(ne) əṭ-ṭōrči qəddām əl-baḥar
PROX.PRES.3MSG the-fisher before the-sea
'**Look, here is** the fisher on the seashore.'

3.2 Some observations regarding syntactic features, functions and constraints of the copula in Mesopotamian Arabic

In this section, I will offer some remarks regarding the function of these copulas alongside observations regarding their syntactic features and constraints, as attested in some of the texts documented by Jastrow (1981). Despite the simple syntactic schematic arising from the overview presented above on the basis of Jastrow (1978: 131–141), the situation attested in actual language samples is more complex.

3.2.1 Mardin (City)

In Mardin city, the relative copula also seems to be attested without any predicate following it, referring back to the head noun of the entire construction.

See also the approach taken by Holes (2016: 89–93), who treats the sequence of demonstrative element followed by personal pronouns in Bahraini Arabic as presentatives.

(22) Mardin (City) [Jastrow 1981: 4]
    sane=ye   *santayn=ye*   *ṭallaʽ*
    year=COP.3FSG   years:DU=COP.3FSG   look:IMP.3MSG
    xams   *ʾsnīn=ye*   *ašqay*   **layye**
    five   years:PL=COP.3FSG   how.much   REL.COP.3FSG
    '(Whether) it is one year, two years, or five years or however much **that (it) is**...'

The presentative copulas and particles (see Section 3.1.4 above) are attested presenting undeniable facts, often as point of departure for future action:

(23) Mardin (City) [Jastrow 1981: 8]
    A wise father has hidden gold to benefit his good-for-nothing son after his death. When the son discovers the treasure, his mother tells him:
    kwā   *abū-k*   *kə=ftakár-ək,*
    PROX.PRES.3MSG   father-your:2MSG   PRF[29]=think:PST:3MSG-you:2MSG
    ḥaṭṭ   *ʽaql-ək*   *fə=ṛās-ək,*
    place:IMP.2MSG   intellect-your:2MSG   in=head-your:2MSG
    ṣēr   *ənsān*
    become:IMP.2MSG   human
    '**Look:** your father has thought of you. So put your brains into their place and be a mensch!'

The positive copula does not necessarily attach to the main or first element of complex predicates but attaches to the element that is contextually the most salient.

(24) Mardin (City) [Jastrow 1981: 10]
    The impoverished son has found his father's gold and invites his false friends to take revenge. He tells them:
    əl-masa   *kə́llət-kən*   *madʽīyīn*   **ʽənd-i=əntən**
    the-evening   all-your:2MPL   invited:MPL   **at-my=COP.2MPL**
    'Tonight, **it's my place** that all of you are invited to.' (lit. 'Tonight all of you invited at my place you are')

---

29 The Arabic dialects of Kurdistan like many other varieties of Arabic employ a number of preverbal particles that are placed in front of inflected present and past tense verbs to express progressivity, intent/futurity, habitual past, pluperfect and perfect. In Mardin, *kə(l)* before the past tense expresses the present perfect.

3.2.2 Qarṭmīn

In Qarṭmīn, nominal clauses can occur under certain conditions also without a copula attaching to the predicate. Such forms, e.g., occur in the apodosis of conditional constructions.

(25) Qarṭmīn [Jastrow 1981: 123]
w   iḏa   mā   rafaʿt   šarṭ-i,   ənt
and if   NEG   lift:PST.2MSG   condition-my   you:2MSG
fə=bayt-ək=∅   w   ana   fə=bayt-i=∅
in=house-your:2MSG=∅   and   I   in=house-my=∅
'If you do not fulfill my condition, you **shall stay** in your house and I **will stay** in mine.'

Moreover, also in circumstantial clauses, the sentence is formed by a juxtaposition of independent pronoun and predicate.

(26) Qarṭmīn [Jastrow 1981: 145]
ʾftakartu   lə=ḥāl-i   kēf   al-bənt   əl-masīḥīyə,
think:PST.1SG   to=self-my   how   the-girl   the-Christian:FSG
w   hīye   āg   ḥəlwe=∅   bə=mītayn w xamsīn   waraqa
and   she   so   pretty:FSG=∅   by=200 & 50   lira
'I thought to myself, how can this Christian girl, since she is so pretty, be worth only 250 liras?'

Also, when a nominal sentence functions as object clause of a verb of perception, it seems that the positive copula is not used, as evidenced by the following minimal pair from the same short story:

(27) Qarṭmīn [Jastrow 1981: 148]
The narrator is waiting in a room; a girl whose mother wants to marry her to him enters. He understands the girl was sent by her mother:
ṭallaʿtu   hīye   zġayyre
see:PST.1SG   she   young:FSG
'I saw (that) she was young.'

(28) Qarṭmīn [Jastrow 1981: 148]
When the mother comes to inquire what he thought of her daughter, the narrator replies:
əl-bənt   zġayyre=ye
the-girl   young:FSG=COP.3FSG
'The girl is (too) young!'

In negative nominal clauses without explicit nominal subject, it is not necessary to have an independent personal pronoun filling the subject slot. Such pronouns, however, can occur as topicalizations.

(29) Qarṭmīn [Jastrow 1981: 122]
mzawwağe=(ə)nti?   manti   mzawwağe?
married:FSG=COP.2FSG NEG.COP.2FSG married:FSG
*tqəl   wallāh **ana** **mana**   mzawwağe*
say:PRS.3FSG   egad   I   NEG.COP.1SG   married:FSG
'(He said:) Are you married or not? She replies: By God, **as for me, I am not** married.' (lit. 'You (are) married? You (are) not married? She says: By God, I, I (am) not married.')

Also, it seems that in negative clauses, the use of the negative copula is not obligatory. The following near-minimal pair is attested in adjacent sentences in the same story with a functional difference:[30]

(30) Qarṭmīn [Jastrow 1981: 145]
A Muslim visitor tells a Christian girl that has just received some suitors, that among Muslims she might have gotten ten or fifteen thousand liras as bride price. He says:
bə=ḥaqq   baqara,   abū-ki   yəḥtī-ki.
by=price   cow   father-your:FSG   give:PRS.3MSG-you:FSG
mū   ḥarām?
NEG.COP.3MSG forbidden?
'Your father is giving you away for the price of a (single) cow! Isn't that a shame?!'

(31) Qarṭmīn [Jastrow 1981: 145])
The girl replies, referring to the bride price common among Muslims:
*layš?   āk   ən-naqəd   mō   **ḥarām**=we?*
why?   that the-bride price   NEG   **forbidden**=COP.3MSG
'Why? Isn't that bride price **forbidden**?'

The impression given by the last example is that the construction without the negative copula is used to focus on the predicate as particularly salient. This could indicate that the negative copula is used primarily to negate the predicative relationship as such, whereas the use of the negating particle allows the

---

30 The negative copula in the first example is a contracted short form of *mawwe*: *mā-huwwā > mawwe > mū. The simple negative particle *mō* in the second example is otherwise attested with present tense verb forms and negates the general and actual present as well as the future.

focus to be shifted to some element other than the predicative relationship. This is also borne out by the following example:

(32) Qarṭmīn [Jastrow 1981: 150]
bayn   ən-nəswān   mō   ǧāyəz=we
between   the-women   NEG   **permitted**=COP.3MSG
ta=ʾabqa   waḥd-i
FUT=remain:PRS.1SG   alone-my
'It is not **proper** for me to remain alone among the women.'

In Qarṭmīn, demonstrative copulas can co-occur together with the positive copula, as in the following example. Here it seems that the entire clause is marked as contextually salient:

(33) Qarṭmīn [Jastrow 1981: 152]
mādām   kī   əl-bent   rāḍye=ye
while   DEM.COP.3FSG   the-girl   **willing:3FSG=COP.3FSG**
'As long as the girl **agrees**, ...'

3.2.3 Bəḥzāni

In Bəḥzāni, a conspicuous usage of the demonstrative copula is its occurrence (often together with the particle *yaʿnī* 'that is', 'that means') before an adjectival attribute as a means to specify the attribute as particularly salient:

(34) Bəḥzāni [Jastrow 1981: 376]
Bəḥzāni,   ẓēʿa,   kī   zġayyġi
Bəḥzāni   village   DEM.COP.3FSG   little:3FSG
'Bəḥzāni is a **really small** village.'

(35) Bəḥzāni [Jastrow 1981: 376]
kəll-a   məǧ=ǧaṣṣ   w   ḥaǧaǧ,   w   bināyāt,
all-her   from=chalk   and   stone   and   buildings:FPL
yaʿnī   kī   malīḥa   ksīġ
that-is   DEM.COP.3FSG   good:3FSG   very
'All (the houses) are made of chalk and stone, and (are) **indeed** very **pretty** buildings.'

In the last example, the entire string from *bināyāti* till *ksīġ* consists of a complex predicate. Note that, as mentioned above, adjectives and pronouns whose referents are inanimate plural nouns, take a feminine singular form. Thus, *bināyāt* is plural, while its attribute *malīḥa* is feminine singular. Likewise, the feminine singular possessive suffix in *kəll-a* refers back to the plural phrase *byūt lāt ẓē'ətna* 'the houses of our village' in the preceding clause in the monologue.

4 Target phrases and their positions

In the dialects of Kurdistan, Targets, i.e., recipients of verbs of giving and taking, addressees of verbs of speaking, as well as other forms of indirect objects are usually expressed by forms of the preposition 'to/for'.[31] This preposition is also used to mark the possessor in possessive statements.

As is the case regarding other prepositions, the form taken before a noun may differ from the form used before a suffixed pronoun (Jastrow 1973: 96). In addition, a third allomorph must be distinguished which occurs directly suffixed to the finite verb expressing the Target. The following cases can thus be distinguished:

(36) Mardin [Jastrow 1981: 22]
ğā xabaṛ **la=l-qəṛāḷ.**
come:PST.3MSG message **to=the-king**
'Word was brought **to the king**.'

(37) Mardin [Jastrow 1981: 6]
mā kəl=kān **ləhu** *ġayr əbən wēḥəd*
NEG PRF=be:PST.3MSG **T.3MSG** except son single
'He only **had** a single son.' (lit. 'There wasn't **to him** but a single son.')

(38) Mardin [Jastrow 1981: 6]
*baqā-***lu** *ōḏāye kama hawne*
remain:PST-3MSG-**T.3MSG** little room like this
'(Only) a small room like this one here remained **for him**.'

That the form of the Target phrase suffixed to a finite verb must be understood as a real suffix is shown not only by its contracted form *lu* '(to/for) him' as in (38) when compared with the non-suffixed pronominal form *ləhu* 'to him' in (37), but also by the fact that it precedes the forms of the direct object pronoun and thus becomes incorporated into the synthetic verb phrase:[32]

---

31 For Targets and their definition, also see Asadpour (this volume, Section 1) and Jügel (this volume, Section 1).
32 Marking the Target of *'ata* 'to give' with a direct objects suffix and not a form derived from the preposition 'to/for' is discussed in the following section.

(39) Mardin [Jastrow 1981: 8]
    awwal mā   (a)ṛaw-hu   mən   bʿīd
    as soon as see:PST.3PL-DO.3MSG from   far
    'as soon as they saw him from far'

(40) Āzəx [Jastrow 1981: 200]³³
    īwzəl-**lu**-wē
    weight:PRS.3MSG-**T.3MSG**-DO.3MSG
    'He weighs it **for him**.'

(41) Mardin [Jastrow 1981: 16]
    w   ʿaṭā-**hu**-we
    and   give:PST.3MSG-**T.3MSG**-DO.3MSG
    'And he gave it **to him**.'

Moreover, in these instances, the forms of the direct object pronouns are homonymous with the forms of the enclitic copula (see also below).³⁴ This indicates that the pronominalized Target has become an incorporated, actual type of object. Incorporation of the Target occurs also in other languages of the region, such as Kurdish (Mukri), Middle Iranian and North-Eastern Neo-Aramaic.³⁵

This constitutes a typologically interesting development, since Semitic languages generally only permit direct, transitive objects to be expressed by suffixed object pronouns on the finite verb itself. Given the close juncture between finite verb and direct object pronouns, the existence of pronominal forms for Targets that take the position between the finite verb and the direct object is highly remarkable.

It is therefore interesting to note that Akkadian in its different varieties developed a distinct "dative" pronoun taking either the form of an independent pronoun following the preposition *ana* 'to/for' or of a pronominal suffix on the verb (Hasselbach-Andree 2019: 100f.).

With verbs of speaking, the preposition 'to/for' is likewise used to introduce the Target:

---

33 Unfortunately, no cases of a 3MSG Target *lu* followed by a 3MSG direct object *we* seem to be attested for the Mardin dialects in the corpus of Jastrow (1981)
34 For a similar phenomenon attested in North-Eastern Neo-Aramaic, see Khan (2012: 115f.)
35 See Asadpour (this volume, Section 2), Jügel (this volume, Section 2.2) and Khan (2012: 115f.).

(42) Mardin [Jastrow 1981: 4]
  qāl   **la=maṛát-u**
  say:PST.3MSG **to=wife-his**
  'He said **to his wife** ...'

(43) Mardin [Jastrow 1981: 13]
  ana *waxt* *la=qəltū-**lkən***
  I time REL=say:PST.1SG-**T.2PL**
  əl-kalb *yākəl*   *laban*
  the-dog eat:PRS.3MSG yoghurt
  'Back when I told **you** the dog eats yoghurt ...'

In addition to the preposition 'to/for', other prepositions such as *ʿal(a)* 'on (top of)/above' and *fə/fi* 'in' can occur in Target phrases in the sense that the verbs in question may be considered semantically to "Target an object" as evident by the parallel attestation of such verbs with a Target introduced by 'to/for'. Moreover, such prepositions may appear as the directive or goal of verbs of movement, often interchangeably with Targets marked by 'to/for'. The following examples exemplify these usages, although they do not form part of the subsequent analysis and discussion:

(44) Mardin [Jastrow 1981: 4]
  The father filled the entire ceiling with gold.
  w *ḥatt-**lu***   *ḥalaqa* *fə=nəṣṣ* *əl-lōḍa.*[36]
  and place:PST.3MSG-**T.3MSG** ring in=middle the-room
  'And he attached **to it** (=the ceiling) a ring in the middle of the room.'

(45) Mardin [Jastrow 1981: 4]
  ḥatt   *xaṭīye* ***fə=raqbə́t-u***
  place:PST.3MSG sin   **in=neck-his**
  'He placed a sin **on his neck**.'

(46) Mardin [Jastrow 1981: 12]
  w *yṛōḥūn* ***la=byūt-en.***
  and go:PST.3PL **to=houses-their**
  'And they go **to their houses**.'

---

[36] In nouns with vowel onsets, the prefixed definite article often causes gemination of the consonant to preserve the syllable structure. The basic form of the noun in question is actually *ōḍa*.

(47) Mardin [Jastrow 1981: 20]
ṛāḥ   'a=l-bayt
go:PST.1MSG on=the-house
'He went **home**.'

In conclusion, Targets in these dialects occur in two positions: (a) postverbal with a nominal Target following a proclitic preposition, usually corresponding to Classical Arabic *li-* 'to/for',[37] and (b) incorporated into the verb phrase in the form of a suffix.[38] This Target suffix, while historically derived from the preposition 'to/for' followed by a possessive suffix, has become truly incorporated as evidenced by its taking the first slot in an object string consisting of Target and direct object suffixes[39] and by the fact that the pronominalized forms of non-Target phrases using the preposition 'to/for' synchronically still consist of the preposition followed by a possessive suffix.[40]

4.1 Targets denoting recipients of giving and taking

In this section, I will discuss the attested features of three verbs, *'aṭa/a'ṭa* 'to give', *ǧāb* 'to bring', and *axaḏ*[41] 'to take' with regard to Target marking.

4.1.1 Targets of 'to give'

As noted above, the verb used in most *qəltu* dialects to denote 'giving', *'aṭa/a'ṭa*,[42] can be constructed with a direct object marking the Target, not the preposition 'to/for'. This, however, is only true regarding a pronominal Target, as in the following example:

(48) Mardin [Jastrow 1981: 18]
kəll   ᵊnhāṛ=ze   ta=tə'ṭī-**ni**
every   day=too   FUT=give:PRS.2MSG-**T.1SG**

---

37 See examples (36, 42, 45, 46, 47) above.
38 See examples (38, 40, 41, 43, 44) above.
39 As in (40, 41) above.
40 As in (37) above.
41 A number of variant forms displaying various sound changes are attested for this verb, e.g. *axaz, axad, axav, aġəz,* etc. (Jastrow 1978: 154f.).
42 Some dialects, e.g., Siirt and Ka'bīye, display a sound shift ' > ḥ, giving a verb *ḥaṭa.*

	xams	lērāt	ḏahab
	five	liras	gold

'Each and every day you shall give me five liras in gold.'

The enclitic particle =ze is usually translated as 'also, too'. However, it serves also (or even more commonly) as a focus or prominence marker and plays an important role in the overall structuring of information. This focusing function is also known from Kurdish, from where this particle is borrowed. The same is true for additional particles meaning 'also, too' used in other dialects such as =zade, =zde in Mḥallami, =zəd in Rīš, or =məne, =əmme in Siirt (Jastrow 1981: XIII).

As mentioned above, when both, the Target and the direct object are pronominalized, the 3rd person direct object is expressed not by its usual direct object suffix but rather by a form that is homonymous to the enclitic copula; this is the case in almost all *qəltu* dialects.[43] In Dēr iz-Zōr, however, an independent object pronoun related to Classical Arabic *iyya-* followed by a possessive suffix is found:

(49) Mardin [Jastrow 1981: 16]
w 'aṭā-**hu**-we
and give:PST.3MSG-**T.3MSG**-DO.3MSG
'And he gave it **to him**.'

(50) Dēr iz-Zōr [Jastrow 1981: 434]
bass il-walad ana mā a'tī-**k** iyyē
just the-boy I NEG give:PRS.1SG-**T.2MSG** DO.3MSG
'As for the boy, however, I will not give him **to you**.'

Although most dialects use the direct object pronouns to mark Targets of the verb 'to give', the dialect of Kaʿbīye features instead a form derived from the preposition 'to/for':[44]

(51) Kaʿbīye [Jastrow 1981: 322]
təḥṭaw-**lna** ādī l-bənt
give:PRS.2PL-**T.1PL** this:FSG the-girl

---

43 Compare the situation in North-Eastern Neo-Aramaic (Khan 2012: 115f.). Interestingly, Classical Arabic allows two object suffixes on a finite verb in ditransitive constructions. In the case of verbs of giving, where the recipient/Target is marked as a direct object, both the recipient and the object of giving would be marked by the normal object suffixes. The fact that *qəltu* dialects mark the direct object (of giving) with a form that is homonymous to the enclitic copula indicates that the Target though formally marked as a direct object has notionally become a true Target.
44 Unfortunately, I was unable to find any examples showing both the Target and the direct object in pronominal form.

'You (shall) give **us** this girl.'

(52) Kaʿbīye [Jastrow 1981: 326]
t=naḥtī-**lken** ṭāyīn
FUT=give:PRS.1PL-**T.2PL** rations
'We shall give **you** food rations.'

While the pronominal Target is thus only rarely marked by a form derived from the preposition 'to/far', nominal Targets generally feature this preposition:

(53) Arbəl [Jastrow 1981: 92]
ʿatáw-ən **lə=ḥāk** zalamə(t)
give:PST.3PL-DO.3PL **to=that:MSG** man
'They gave them **to that man**'

(54) Arbəl [Jastrow 1981: 94]
t=aʿti **l=axū-y** karm-i
FUT=give:PRS.1SG **to=brother-my** vineyard-my
'I will give my vineyard **to my brother**.'

(55) Mardin [Jastrow 1981: 22]
ʿaṭaw **la=l-qáyəġči** waraqatayn tāte
give:PST.3PL **to=the-boatswain** two pounds three
'They gave two, three pounds **to the boatswain**.'

As far as can be told from the corpus, there is no free word order variation in the sense that either the direct object or the Target can stand in front of the verb marking the focus.[45] Likewise, no variation in the order of nominal Target and nominal object is found. Rather, the constituent order is determined by two factors: pronominal forms precede nominal ones, and, if both are either nominal or pronominal, Target phrases are given precedence. This means, pronominal Targets precede pronominal as well as nominal (direct) objects, pronominal objects precede nominal Targets, and nominal Targets precede nominal objects:

a) verb-Target[pron]-object[pron]
b) verb-Target[pron] object[nom]
c) verb-object[pron] Target[nom]
d) verb Target[nom] object[nom]

---

[45] Although I am currently unable to adduce examples for one of the two adjuncts occurring as frontal topicalization before the finite verb, they probably do occur, since they are attested for the other, semantically similar verbs of giving and taking (see below).

Focus marking, it seems, is instead achieved with the help of particles such as the above mentioned =ze:

(56) Mardin [Jastrow 1981: 14]
 t=aʻtī-k=ze   *ḥmēlət-ək*
 FUT=**give:PST.1SG-2MSG**=FOC porterage-your:MSG
 '**Don't doubt that I shall give you** your porterage.' (lit. 'I shall give you, too, your porterage')

The interlocutor in this narrative is placed more than once in focus, while at the same time, there are no other possible referents which would warrant the use of =ze in the sense of 'also, too'. This shows that =ze does not mark a contrastive focus, but rather highlights the verbal phrase and marks the assuredness of the action as salient.

There is, however, a single example, where the Target occurs before the finite verb. However, I am at present unsure how to interpret this case, although there may be some degree of focus on the Target. The attestation occurs in a story on the destruction of Kaʻbīye:

(57) Kaʻbīye [Jastrow 1981: 334]
 The village priest is helping his flock escape the massacre:
 šā=kəll   *wēḥəd-en*   *ḥata*   *ṣōmūnāy*
 to=all **single-their** give:PST.3MSG breadroll
 '**To every single one of them** he gave a breadroll.'

This preposition is attested also elsewhere in Kaʻbīye as Target and indirect object marker. It is a development of the phrase *li-šaʼn* 'for the purpose of', which has undergone phonetic reduction and grammaticalization in this dialect (Jastrow 1981: 166 fn. 17).

4.1.2 Targets of 'to bring'

The main verb in the Arabic dialects of Kurdistan used to denote 'bringing' is *ǧāb*. It occurs mainly without a Target, unlike the verb 'to give', where a recipient is almost always necessary and at least contextually understood.[46] The verb 'to

---

46 This is possibly due to the semantic history of this verb. Different verbs denoting 'to come' such as *ǧāʼ* and *atā* have shifted semantically to denote 'to bring' when constructed with the instrumental preposition *bi-* 'in, with'. In many dialects, this semantic shift has resulted in a

bring' can, however, be constructed with an optional Target denoting the direction or destination of the act of bringing, since the verb 'to bring' in itself only indicates that the location of an object is being changed. Often, the destination is contextually identified as the place where the subsequent actions take place:

(58) Rīš [Jastrow 1981: 78]
 baʿd lə=ğā̀bu ∅ š-šaxəṛ
 after REL=bring:PST.3PL ∅ the-bearer frame
 t=īğībūn ∅ ʿəwā̀qīl=zəd
 FUT=bring:PRS.3PL ∅ wooden angle finders=too
 'After bringing the bearer frame (to the workplace), they will also bring the wooden angle finders (there).'

Regarding the variation in the position of the Target and the direct object, the same rules that were established in the previous section, seem to apply, although not all cases appear to be attested:[47]

(59) Mardin [Jastrow 1981: 30]
 yətlawn kīsət mağīdīye
 fill:PRS.3PL bag Mağīdīs
 *w yğībū-**le**-ye*
 and bring:PRS.3PL-**T.3MSG**-DO.3FSG
 'They fill a bag with Mağīdīs and give it **to him**.'

(60) Mardin [Jastrow 1981: 40]
 *yğībū-**lu** məxēməl mən barra*
 bring:PRS.3MSG-**T.3MSG** towels from outside
 'They bring **him** towels from outside.'

(61) Siirt (town) [Jastrow 1981: 250]
 *awle mā nğib-u **l=bayt***
 as soon as bring:PRS.1PL-DO.3MSG **to=house**
 'as soon as we bring it (=the goat hair) **home**.'

(62) Kəndērīb [Jastrow 1981: 62]
 w əl-ʿaṛabāṇa ᵊğğīb-a lə=d-dēʿa
 and the-vehicle bring:PRS.3MSG-DO.3FSG **to=the-village**
 'And the vehicle brings it (=the long stone block) **to the village**.'

---

new verb through a process of univerbation, *ğāʾa bi-X > ğāb X*. On this process and the semantic shift, see Bloch (1991: 41–53).

47 As mentioned above, precedence is given to pronominal forms, evidently due to their lack of stress, and all things being equal, Targets precede the direct objects.

The Target phrase can even be reflexive or indicate a beneficiary:[48]

(63) Qarṭmīn [Jastrow 1981: 134]
 əxti ǧībī-**lki** frāġ
 sister-my bring:IMP.2FSG-**T.2FSG** container
 'My sister, fetch for **yourself** a container!'

Another verb attested frequently with the meaning 'to bring' is *wadda*, literally 'take away, take to a place, fetch'. This verb occurs almost always with a Target phrase denoting the destination. The first of the following examples features the preposition *ša=* 'to' mentioned above, while the second shows that (pro)nominal objects can occur as fronted topics of the verb 'to bring'; they are, however, resumed pronominally in the usual direct object position:

(64) Āzəx [Jastrow 1981: 180]
 waddī-yu ša=l-qəmandāṛ
 bring:IMP.2FSG-DO.3MSG to=the-commander
 'Bring it to the officer!'

(65) Arbəl [Jastrow 1981: 94]
 hāy=zəd waddī-**lək**-ye
 this:FSG=too bring:IMP.2FSG-**T.2FSG**-DO.3MSG
 'Also this one, take it **for yourself**!'

In the last example, the Target again indicates the beneficiary.

The following two examples are very curious. Here both adjuncts occur at the same time in pronominal as well as nominal form. The first example shows a topicalized fronted object that is resumed as object pronoun in the form of the enclitic copula, while the Target seems to be in focus. The second attestation is more difficult. Contextually, it seems that none of the constituents are specifically topicalized or focused, but rather the entire clause as such seems to be marked as salient, presenting some kind of assurance to the addressee:

(66) Āzəx [Jastrow 1981: 206]
 w kəll-ən tə=dǧīb-**lək**-ən **šān-ək**
 and all-their FUT=bring:PRS.3FSG-**T.2MSG**-DO.3PL **to-you**
 'And she will bring all of them **to you**!' (lit. 'And all of them, she will bring you them, to you')

(67) Āzəx [Jastrow 1981: 206]
 w əl-bənt=əst **tə=dǧīb-lək-ən**
 and the-gir=too FUT=**bring:PRS.3FSG-T.2MSG-DO.3PL**

---

48 See also Section 4.1.3 below. For this type of personal datives, see Haddad (2018).

kəllát-ən   šān-ək
all-their   to-you:2MSG
'And the girl **will indeed** bring all of them **to you!**' (lit. 'And the girl too, **she will bring you them**, all of them, to you.')

4.1.3 Targets of 'to take'

The third main verb with a recipient Target is the verb 'to take'. Here, the verb usually does not occur with an explicit Target, as in the first example below. However, on occasion, a reflexive Target phrase derived from the preposition 'to/for' and suffixed to the verb is attested. I am still uncertain regarding the conditions causing the expression of these co-referential datives. Most cases seem to be instances of beneficiaries, although some examples point perhaps towards a perfective nuance accompanying the beneficiary recipient, indicating the completion and non-revocability of the acquisition (as in (69), (71) and (72), possibly).[49] Examples are found most frequently with imperatives:

(68) Mardin [Jastrow 1981: 12]
kəll   wēḥəd   lēra   ḏahab   axaḏu
every   single   pound   gold   take:PST.3PL
'Everyone took a gold pound.'

(69) Bəḥzāni [Jastrow 1981: 404]
kaku   dəst   mən   əz-záhab,
there is pot   from   the-gold
ġōḥ   xəz-lək-ū   w   taʿāl
go:IMP.2MSG take:IMP.2MSG-**T.2MSG**-DO.3MSG and   come:IMP.2MSG
'There is a pot of gold, go, take it **into your possession** and come back!'

(70) Kaʿbīye [Jastrow 1981: 318]
From the narrative on the destruction of Kaʿbīye, describing the atrocities and rape committed by the soldiers:
kā=yṛūḥ   yāxəd-**lu**   maṛa
HAB=go:PRS.3MSG   take:PRS.3MSG-**T.3MSG**   woman
'He would go and take **himself** a woman!'

---

[49] For personal and co-referential datives in Arabic and their pragmatic function(s), see Haddad (2018).

(71) Qarṭmīn [Jastrow 1981: 138]⁵⁰
xəl-**lək** *farás-i*
take:IMP.2MSG-**T.2MSG** horse-my
'Take my horse (**and be off**)!'

(72) Kaʿbīye [Jastrow 1981: 358]
kāġdāy xədū-**lken**
ticket take:IMP.2PL-**T.2PL**
'**Get yourself** a ticket!'

4.2 Targets denoting addressees of verbs of saying

The two main verbs of saying that are attested with Targets are *qāl* 'to say, tell' and *ḥaka* 'to tell, narrate'. The former is used with a Target phrase derived from the preposition 'to/for', whereas the latter utilizes the direct object pronouns to express Targets.

The first example is highly instructive, featuring two attestations of *qāl* 'to say'. The first instance shows the unmarked usage with the Target suffixed to the final verb, whereas in the second one, the Target is pronominalized but precedes the finite verb, signaling focus and displaying an expressive gemination of the pronominal element. The second example likewise has two attestations of the verb of saying. Here, a nominal Target is placed before the verb, denoting contrastive focus with the pronominal Target in the second instance of the verb, while the object first occurs topicalized in front of the finite verb and is resumed by direct object pronouns in both occurrences of the verb 'to say'. The second verb of saying, *ḥaka* 'to tell' behaves very similarly to the verb ʿaṭa/aʿṭa 'to give'.

(73) Mardin [Jastrow 1981: 14]
qāt-lu *qūm* *warak*,
say:PST.3FSG-T.3MSG rise:IMP.2MSG fella
lə-kk *aqūl*
to-your:MSG say:PRS.1SG
'She said: Get up, fella, it's **you** I'm talking to!'

(74) Qarṭmīn [Jastrow 1981: 124]
faqat *ana* *al-masale* *hē* **lə=ʾaḥḥad**
only I the-story this **to=someone**

---

50 Here and in the following example, the final (inter)dental of the verbal root assimilates to the onset of the Target phrase.

mā   qəltū-wa,   t=aqəl-**lek**-ye
NEG   say:PST.1SG-DO.3FSG   FUT=say:PRS.1SG-**T.2MSG**-DO.3FSG
'As for this story, I have not told it **to anybody**, but I want to tell it **to you!**'

(75)   Mardin   [Jastrow 1981: 8]
       ta'áw   t=aḥkī-**kən**   ḥəkkōyət   əl-laban
       come:IMP.2PL   FUT=tell:PRS.1SG-**T.2PL**   story   the-yoghurt
       'Come, I will tell **you** the story of the yoghurt!'

(76)   Mardin   [Jastrow 1981: 10]
       fīyu   ḥəkkōye   t=aḥkī-**kən**-yē
       there is   story   FUT=tell:PRS.1SG-**T.2PL**-DO.3FSG
       'There is a story I want to tell you.'

5 The question of language contact

Language contact has often been mentioned as a possible or likely source for certain features characterizing the Arabic dialects of Mesopotamia. This includes the occurrence of copulas in these dialects (Akkuş 2020: 147ff.). However, the discussion does not generally move beyond the point of stating that these Mesopotamian varieties of Arabic are similar to Kurdish, Turkish and Aramaic in that they, too, use a copula in nominal clauses.

Nevertheless, more extensive congruence exists between some of the characteristics exhibited by Mesopotamian Arabic copulas and the copulas in North-Eastern Neo-Aramaic. Thus, e.g., North-Eastern Neo-Aramaic possesses deictic copulas that are formed by combining deictic elements with the enclitic positive copula (Khan 2012: 112ff.), the function of which is to "draw attention to a referent in the environment" (Khan 2012: 113).

Moreover, both Mesopotamian Arabic and North-Eastern Neo-Aramaic mark the second (pronominal) direct object in ditransitive constructions that have an indirect (Target) as well as a direct object by a pronominal form that is homonymous with the enclitic copula but functions entirely as suffix pronoun (Jastrow 1978: 296ff.; Khan 2012: 115f.), thus aligning both copulas and Targets in these two language groups:

(77)   Christian Qaraqosh   [Khan 2012: 116]
       kewi-**ləḥ**-ilə
       give-**T.3MSG**-DO.3MSG
       'They give him to him.'

(78) Christian Qaraqosh [Khan 2012: 116]
kewi-ləḥ-ila
give-T.3MSG-DO.3FSG
'They give her to him.'

(79) Qarṭmīn [Jastrow 1978: 297]
ʿaṭayt́ū-hu-wē
give:PST.1SG-T.3MSG-DO.3MSG
'I gave him to him.'
ǧəbt́ū-lu-yē
bring:PST.1SG-T.3MSG-DO.3FSG
'I brought her to him.'

In this context, it must be kept in mind that already Syriac, the liturgic language of Eastern Christianity, possesses a copula that developed from personal pronouns in enclitic position and is attached to participles and adjectives (Nöldeke 1880: 42f.). The fact that Christian Baghdadi possesses an enclitic copula, while Jewish Baghdadi does not, is noteworthy.

Interestingly, the morpho-syntactic restructuring in these varieties of Arabic has started to impact verbs like ʿaṭa/aʿṭa which are originally constructed with two direct objects, one denoting the received object, and the other the recipient. This process has reached the point where the first suffix always refers to the Target and the second to the received object, with the object being expressed by a form identical to the copula, similar to verbs with Target forms derived from the preposition 'to/for' followed by a possessive suffix, as in the last example above. Moreover, nominal Targets of ʿaṭa/aʿṭa are marked by the preposition 'to/for' (see also Section 4.1.1 above).

6 Conclusions

The syntax of copulas as well as Targets in the Arabic dialects of Kurdistan is a fascinating topic. Like other aspects of the syntax of Kurdistan Arabic dialects, they remain seriously understudied. Nevertheless, the picture that emerges is that Mesopotamian Arabic dialects that possess copulas align better with Aramaic, especially North-Eastern Neo-Aramaic, than with other languages spoken throughout the contact area.

This also holds true for the suffixation of Targets on the verb and the formal homonymity between the second (direct object) pronoun in ditransitive constructions and the enclitic copula. However, the incorporation of pronominal Targets also represents a certain alignment with some varieties of Kurdish and other Iranian languages.

Acknowledgments: I wish to thank the editors of this volume, Hiwa Asadpour and Thomas Jügel, for the inclusion of my paper and their very many helpful comments and suggestions. I am likewise grateful to the anonymous reviewer for their corrections, comments, suggestions and bibliographic references. Remaining shortcomings are my own responsibility.

Abbreviations

1, 2, 3	1st, 2nd, 3rd person
COP	copula
DEM.COP	demonstrative copula
DIST.PRES	distal presentative
DO	direct object
DU	dual
FOC	focus
FPL	feminine plural
FSG	feminine singular
FUT	futurity (preverbal particle)
HAB	habitual past (preverbal particle)
IMP	imperative
MDU	masculine dual
MPL	masculine plural
MSG	masculine singular
NEG	negator
NEG.COP	negative copula
PL	plural
PRF	perfective (preverbal particle)
PROX.PRES	proximal presentative
PRS	present tense
PRES.PART	presentative particle
PST	past tense
REL	relativizer
REL.COP	relative copula
SG	singular
T	Target

References

Akkuş, Faruk. 2020. Anatolian Arabic. In Christopher Lucas & Stefano Manfredi (eds.), *Arabic and contact-induced change*, 135–158. Berlin: Language Science Press.

Asadpour, Hiwa. this volume. Word order in Mukri Kurdish – the case of incorporated targets.

Behnstedt, Peter. 1992. Qəltu-Dialekte in Ost-Syrien. *Zeitschrift für Arabische Linguistik* 24. 35–59.

Behnstedt, Peter, Wolfdietrich Fischer & Otto Jastrow. 1980. *Handbuch der arabischen Dialekte*. Wiesbaden: Harrassowitz.

Blanc, Haim. 1964. *Communal dialects in Baghdad*. Cambridge, MA: Center for Middle Eastern Studies.

Bloch, Ariel A. 1991. *Studies of Arabic syntax and semantics*. Wiesbaden: Harrassowitz.

Camilleri, Maris & Louisa Sadler. 2019. The grammaticalisation of a copula in vernacular Arabic *Glossa: A Journal of General Linguistics* 4(1). 137. 1–33. DOI: https://doi.org/10.5334/gjgl.915 4(1). 137. DOI: https://doi.org/10.5334/gjgl.915

Fischer, Wolfdietrich. 2001. *A grammar of Classical Arabic*. New Haven & London: Yale University Press.

Goldenberg, Gideon. 1998. On Syriac sentence-structure. In Gideon Goldenberg, *Studies in Semitic linguistics: Selected writings*, 525–568. Jerusalem: Magnes Press.

Goldenberg, Gideon. 2013. *Semitic languages: Features, structures, relations, processes*. Oxford: Oxford University Press.

Haddad, Youssef A. 2018. The pragmatics-syntax division of labor. The case of personal datives in Lebanese Arabic. In Reem Bassiouney & Elabbas Benmamoun (eds.), *The Routledge handbook of Arabic linguistics*, 155–179. London: Routledge.

Hasselbach-Andree, Rebecca. 2019. Akkadian. In John Huehnergard & Na'ama Pat-El (eds.), *The Semitic languages*, 95–116. London: Routledge.

Holes, Clive. 2006–2009. Bahraini Arabic. In Kees Versteegh, Mushira Eid, Alaa Elgibali, Manfred Woidich, & Andrzej Zaborski (eds.), *Encyclopedia of Arabic language and linguistics*, vol. 1, 241–255. Leiden: Brill.

Holes, Clive. 2016. *Dialect, culture, and society in eastern Arabia*. Vol. 3: *Phonology, morphology, syntax, style*. Leiden: Brill.

Ingham, Bruce. 2006–2009. Khuzestan Arabic. In Kees Versteegh, Mushira Eid, Alaa Elgibali, Manfred Woidich, & Andrzej Zaborski (eds.), *Encyclopedia of Arabic language and linguistics*, vol. 2, 571–578. Leiden: Brill.

Jastrow, Otto. 1973. *Daragözü – eine arabische Mundart der Kozluk-Sason-Gruppe (Südostanatolien). Grammatik und Texte*. Nürnberg: Hans Carl.

Jastrow, Otto. 1978. *Die mesopotamisch-arabischen qəltu-Dialekte. Band I: Phonologie und Morphologie*. Wiesbaden: Steiner.

Jastrow, Otto. 1981. *Die mesopotamisch-arabischen qəltu-Dialekte. Band II: Volkskundliche Texte in elf Dialekten*. Wiesbaden: Steiner.

Jastrow, Otto. 2006–2009a. Anatolian Arabic. In Kees Versteegh, Mushira Eid, Alaa Elgibali, Manfred Woidich, & Andrzej Zaborski (eds.), *Encyclopedia of Arabic language and linguistics*, vol. 1, 87–96. Leiden: Brill.

Jastrow, Otto. 2006–2009b. Iraq. In Kees Versteegh, Mushira Eid, Alaa Elgibali, Manfred Woidich, & Andrzej Zaborski (eds.), *Encyclopedia of Arabic language and linguistics*, vol. 2, 414–424. Leiden: Brill.

Jügel, Thomas. this volume. Word order variation in Middle Iranic: Persian, Parthian, Bactrian, and Sogdian.
Khan, Geoffrey Allan. 1997. The Arabic dialect of the Karaite Jews of Hīt. *Zeitschrift für Arabische Linguistik* (34). 53–102.
Khan, Geoffrey Allan. 2006. Some aspects of the copula in North West Semitic. In Steven Ellis Fassberg & Avi Hurvitz (eds.), *Biblical Hebrew in its Northwest Semitic setting: Typological and historical perspectives*, vol. 1, 155–176. Jerusalem: Magnes Press.
Khan, Geoffrey Allan. 2012. Grammaticalization of the copula in North-Eastern Neo-Aramaic. In Domenyk Eades (ed.), *Grammaticalization in Semitic*, 109–125. Oxford: Oxford University Press.
Leitner, Bettina. 2020. Khuzestan Arabic. In Christopher Lucas & Stefano Manfredi (eds.), *Arabic and contact-induced change*, 115–134. Berlin: Language Science Press.
Nöldeke, Theodor. 1880. *Kurzgefasste syrische Grammatik*. Leipzig: Weigel.
Palva, Heikki. 2006–2009. Dialects: Classification. In Kees Versteegh, Mushira Eid, Alaa Elgibali, Manfred Woidich, & Andrzej Zaborski (eds.), *Encyclopedia of Arabic language and linguistics*, vol. 1, 604–613. Leiden: Brill.
Procházka, Stephan. 2020. Arabic in Iraq, Syria, and southern Turkey. In Christopher Lucas & Stefano Manfredi (eds.), *Arabic and contact-induced change*, 83–114. Berlin: Language Science Press.
Reckendorf, Hermann. 1921. *Arabische Syntax*. Heidelberg: Winter.
Talay, Shabo. 2011. Arabic dialects of Mesopotamia. In Stefan Weninger, Geoffrey Allan Khan, Michael P. Streck & Janet C. E. Watson (eds.), *Semitic languages: An international handbook*, 909–920. Berlin: Mouton.
Versteegh, Kees. 1997. *The Arabic language*. New York: Columbia University Press.
Versteegh, Kees. 2014. *The Arabic language*. Edinburgh: Edinburgh University Press.
Walter, Mary Ann. 2020. Cypriot Maronite Arabic. In Christopher Lucas & Stefano Manfredi (eds.), *Arabic and contact-induced change*, 157–174. Berlin: Language Science Press.
Watson, Janet C.E. 2011. Arabic dialects. In Stefan Weninger, Geoffrey Allan Khan, Michael P. Streck, & Janet C. E. Watson (eds.), *Semitic languages: An international handbook*, 851–896. Berlin: Mouton.
Zemer, Hila. 2008. The copula in the Christian Arabic dialect of Baghdad – possible origins. In Stephan Procházka & Veronika Ritt-Benmimoun (eds.), *Between the Atlantic and the Indian oceans: Proceedings of the 7th AIDA conference*, 505–514. Wien: LIT.

Paul M. Noorlander and Dorota Molin
Word order typology in North-Eastern Neo-Aramaic

Towards a corpus-based approach

Abstract: This study applies a corpus-based quantitative approach to the word order typology in two distinct Jewish North-Eastern Neo-Aramaic dialects: Sanandaj representing an OV type and Dohok a VO type. Despite the difference in the default position of the object, these two dialects share the clause-final position of Goals. Still, Goals and other obliques are more flexible in Sanandaj, while objects are more flexible in Dohok. An argument in narrow focus (any in Sanandaj and object in Dohok) is typically immediately preverbal in both. The position of objects outside predicate or narrow focus is typically inversed.

Keywords: ditransitive clauses; goal placement; Modern Aramaic; object-verb; word order change

1 Towards a NENA corpus

North-Eastern Neo-Aramaic (NENA) is an umbrella term subsuming highly diverse dialects of Jewish (J.) and Christian (C.) minorities scattered across the globe. Their original territory comprised parts of South East Turkey, North Iraq and West Iran. NENA is closely related to the Neo-Aramaic of Christians of Ṭur ʿAbdin (Ṭuroyo, SE Turkey, NE Syria). Being by daily necessity at least bilingual, their dialects were influenced by neighboring Iranian (Kurdish, Persian), Turkic, Armenian and Arabic dialects (e.g., Khan 2008b; Noorlander 2014). The Greater Zab River functions as an isogloss dividing Jewish dialects into two major groups: *lishana deni* (LDJ) in the west and the Trans-Zab Jewish group (TZJ) in the East, while the dialects around the settlement Barzan represent a transition zone (Mutzafi 2008b). The *lishana deni* dialects such as Dohok, Zaxo

---

Paul M. Noorlander, Faculty of Asian and Middle Eastern Studies, University of Cambridge, Sidgwick Avenue, CB3 9DA Cambridge, United Kingdom, E-mail: pmn32@cam.ac.uk
Dorota Molin, Faculty of Asian and Middle Eastern Studies, University of Cambridge, Sidgwick Avenue, CB3 9DA Cambridge, United Kingdom, E-mail: dm605@cam.ac.uk

and Amedia are closer to the local Christian dialects. The Trans-Zab group has been heavily influenced by contiguous Iranian languages (cf. Noorlander 2014).

NENA morphosyntax is considerably complex and lies beyond the scope of this article[1]. All NENA varieties use differential object marking (DOM) strategies and are characterized by a constructional split between imperfective and perfective clauses and show distinct morphological alignment types in their verbal person marking. Historically, this split emerged out of the active (*šāqel- 'taking') and resultative participle (*šqīl- 'taken'). Dialects show distinct morphological alignment types in their verbal person marking.

Dialects also differ in word order typology. Word order is said to be relatively flexible and usually varies depending on the discourse in Neo-Aramaic (e.g., Hoberman 1989: 100). The TZJ varieties, however, are generally described as OV, while LDJ dialects are VO. The OV order is presumably the result of contact with the neighboring OV-dominant languages. The preservation of the earlier Aramaic VO order in the LDJ dialects indicates a greater degree of resistance to cross-linguistic interference relatively to the TZJ group. The recent years recorded a growing interest in word order preferences for NENA dialects from an areal perspective (e.g., Haig 2014, to appear; Stilo 2018), focusing on arguments which tend to occur postverbally in the OV type. Stilo (2018) studied this for C. Barwar, J. Zakho (i.e., LDJ), J. Urmi and J. Sanandaj.[2] Goal-like arguments are commonly expressed postverbally in the Iranian languages of Northern Iraq and Western Iran. The VG order is also commonplace in both the OV and VO types of NENA. According to Haig, therefore, the VG order in the Iranian languages of that area may in itself reflect the 'ancient imprint of Aramaic' (Haig 2014: 422; cf. Stilo 2018).

To our knowledge, however, no quantitative study of the basic word order in NENA has been conducted to-date. Statistical corpus-based research supporting grammatical description still remains a desideratum for NENA. Our aim is to contribute to the areal word order typology by comparing the western (e.g., LDJ) and eastern (e.g., TZJ) Jewish NENA dialect types using two case studies from a corpus-based approach. The study of frequency-based effects requires an annotated corpus. Since such a corpus is still under development, our findings remain preliminary.

The NENA varieties studied are: the dialect of Dohok (NW Iraq) representing LDJ, and of Sanandaj (W Iran) representing TZJ. The Dohok corpus of ca. 6,000

---

[1] For overviews, see Khan (2011), Coghill (2016: 55–101) and Noorlander (2021).
[2] His NENA data are based on texts from Khan (2008a–b and 2009), and presumably from Meehan and Alon (1979) for J. Zakho.

words is based on Molin's fieldwork with five informants. It consists of personal narratives, descriptions of past customs and folktales. The first description of Dohok is now available in Molin's doctoral thesis.[3] This dialect was historically spoken by the Jews of the town of Dohok (Duhok) and is today preserved by a small community of immigrants from Dohok living in and around Jerusalem (Israel). In contrast to LDJ dialects such as Zakho and in parallel to Betanure, Dohok is characterized by a conservative phonology in which the earlier-Aramaic interdentals are still preserved. In morphosyntax, a noteworthy feature of Dohok is that the strategy of marking objects internally on the past base is virtually obsolete, with the *qam*-based past transitive paradigm (corresponding to, for instance, *kəm-* and *qəm-* elsewhere in NENA) functioning in its stead. In many other areas, however, including word order, the Dohok dialect bears a close resemblance to the entire LDJ cluster from the area.

The Sanandaj corpus of ca. 7,600 words is based on Khan (2009, excl. poem E) with four informants. It consists of spontaneous spoken language about past customs, anecdotes and life stories, as well as folktales. Table 1 below shows the counts of annotated verbal clauses per dialect. The second row represents the percentage of verbal clauses with overt independent arguments (NPs and independent pronouns).

Table 1: Total counts of annotated data.

	Dohok	Sanandaj
Total verbal clauses	1,577	1,742
of which with full (pro)nominals	44%	70%

2 Methodology

The two corpora were tagged for transitivity and argument structure. The value of the transitivity tag ('intransitive, 'transitive' and 'ditransitive') was established lexically and not on the basis of explicit argument marking in the clause. The handful of verbs that are ambivalent were marked on the basis of the usage. For example, a verb like 'make' in NENA can be transitive ('make bread') or ditransitive ('make grains into flour'; see ex. 14g and 14i in Section 6.2).

---

3 See Molin (2021).

Following Andrews (2007), our argument class annotation distinguishes between the S(ubject) of intransitive clauses and the A(gent)-like subject and P(atient)-like object of transitive verbs. We also distinguish between T(hemes) and R(ecipients) for ditransitives. When discussing objects in general, however, we subsume both T and P under the category of O. This is because it is generally T, not R, which triggers agreement like the P in prepositional indirect object constructions, which is the most common ditransitive construction in Aramaic. When R triggers agreement, however, R is considered the primary object.[4] The category G(oal) (referred to as 'Target' elsewhere in this volume) subsumes in NENA: all prepositional Goal-like arguments expressing the destination of movement verbs and of caused motion (like 'put'), the Recipient of ditransitive verbs like 'give' and 'show', the addressee of 'say, tell', 'speak' and 'ask', beneficiaries and the patient-like complement of 'look at'. Occasionally, the preposition marking of such Goals is dropped, cf. (1a) and (1b) below, in which case it was still subsumed under Goals. The non-prepositional recipient-like argument in double object constructions, however, has been treated distinctly from G.

(1) Sanandaj [Khan 2009: 464, B:74]
 a. k-e-x-wa-o ∅-bela
 b. k-e-x-wa-o baqa-bela
  IND-come.IPFV-S:1PL-PST back-to house
  'We would come back home.'

Finally, TZJ uses a number of complex predicates, particularly with ʔwl 'do' and dry 'throw'. The nominal elements of such predicates were not included as objects in our study, unless they trigger agreement as in (1c).

(1) c. hičkas ay-ḥašta wil-a-wa-le
  nobody DEM-work.FS do.PFV-P.3FS-PST-A.3MS
  'Nobody had done such a thing.' [Khan 2009: 414, A:25]

Statistical significance tests were used to demonstrate the significance of frequencies. The Chi square test has been used when the total number of observations was above 20, and none of the expected frequencies were below 5. Otherwise, the Fisher exact test was used (Levshina 2015: 213–14). The value of degrees of freedom is 1, unless stated otherwise. Extremely significant probability-values were kept at p<0.0001. In tables listing frequencies, expected frequencies are given in parentheses.

---

4 See further Section 6 and Noorlander (2018: 144–174, 395–402) for more details. Ditransitive alignment depends on various factors, including the criteria selected to determine this.

3 General frequencies

3.1 A, S and P: basic differences

The basic differences between S, A (including ditransitive clauses) and P are in the frequency of their full expression in the clause and in the part of speech used to express them (NP, independent pronoun, pronominal affix), as displayed in Table 2 below. Since cross-indexing of S and A co-nominals is obligatory in NENA (see Noorlander 2021: 74, 78, 95), only P can be left unexpressed. For both dialects, the difference in the frequency of full expression of S and A is statistically significant ($p<0.0001$). A is realized fully least frequently, with S being more frequent and P being most frequent. Moreover, P is much more likely to be expressed by an NP, while A is least likely to be expressed in this way. This is consistent with cross-linguistic corpus data (Haig and Schnell 2016).

Table 2: Argument class frequency per argument type.

	Dohok (LDJ)			Sanandaj (TZJ)		
	A	S	P	A	S	P
Total verbal clauses	1039	538	614	1223	524	700
of which full NP	10%	24%	47%	11%	28%	55%
of which independent PRO	8%	6%	0.4%	7%	10%	3%
of which affixal PRO			25%			15%

LDJ dialects prefer to mark pronominal objects as verbal affixes. By contrast, it is typical of but not exclusive to the TZJ varieties to have distinct series of independent object pronouns. The basis of such pronouns is the preposition ʔəl (often reduced to l-) and/or the particle did-. The ʔəl-based pronominal objects always follow the perfective past verb in the Sanandaj corpus. When postverbal, however, the ʔəl-series tends to cliticize to the preceding verbal form, e.g., graš-li ʔalaf → graš-li-laf 'I pulled her' (cf. Khan 2009: 158).[5] This unmarked postverbal position for pronominal objects in Sanandaj against preverbal full NPs is typologically unusual and presumably points to the original AVO structure of this language. We have excluded this ʔəl-series from the following discussion.

---

[5] See Noorlander (2021: 103–105, 270–275, 367, 369) for possible motivations for this.

3.2 Default word order and flexibility

Only 8.5% of the verbal clauses in Dohok and 7.3% in Sanandaj contain two arguments fully expressed. The expected order of such two argument clauses is fixed to a significant degree. Table 3 below shows the permutations that occur for A followed by O and A or A followed by G relative to verb-initial, medial and final positions. The "Other" row includes other permutations like OAV, OVA or GVS etc. Dohok is consistently verb-medial: in ca. 92% (122/133) cases, S/A are preverbal, O/G postverbal. Thus when two full nominal arguments are expressed, verb-final order is only a clear preference for monotransitives in Sanandaj, and is virtually non-existent in Dohok.

Table 3: Overall word order typology.

	Dohok (LDJ)			Sanandaj (TZJ)		
	V-initial	V-medial	V-final	V-initial	V-medial	V-final
AO	0	69/70 (98.5%)	1/70	0	7/86	79/86 (92%)
SG/AG	5/57	53/57 (93%)	0	0	89/100 (89%)	11/100
Other	4	0	2	0	3	13

Both dialects prefer the order VG for clauses with full S or A in general (Dohok 93%, Sanandaj 89%). Thus, such Goals even follow the verb in TZJ dialects, which are typically AOV. This postverbal placement in TZJ (Sanandaj) is illustrated below by Goals such as the addressee of the verb *ʔmr* 'say' in (2a) and the Recipient of *ywl* 'give' in the indirect preposition construction in (2b).

(2)   Sanandaj   [Khan 2009: 436, A:107]
    a. *baxt-ef*   *mi-ra*   **baqa-gor-ăkè...**
       woman.FS-his say.PFV-A.3FS to-man.MS-DEF
       'His wife said **to the man...**'
    b. *pul-e*   *ṭălabkār* *k-w-i-wa-le-o*   **baq-ù**
       money-of creditors IND-give.IPFV-A.3PL-PST-T.3PL-back to-them
       'They would give the creditors' money back **to them.**' (440, B:9)

As expected for OV languages (Siewierska 1998: 493), verb-first does not occur in Sanandaj and AVO (7/202) occurs besides OAV (13/202); the OVA (3/202) variant is relatively infrequent among OV languages and thus somewhat unexpected for Sanandaj.

At face value, these findings seem to suggest word order is slightly more flexible in Sanandaj. However, they only pertain to clauses with two arguments fully expressed. Arguments are more likely to be expressed postverbally in general in Dohok (65% against 35%) but more likely preverbally in Sanandaj (69% against 31%). The preferences, however, naturally differ per argument class S, A, O, G) and argument type. While OV and VA do occur in Dohok (see further below), they have not been attested together (**OVA). Similarly, VO and VA order are possible variants in Sanandaj, but they do not occur together (e.g., **VAO/VOA).

Table 4 below shows that the preverbal placement of the A is strongly preferred in both dialects. Nominal Ps are expressed postverbally in Sanandaj in only 4.4% of the cases, and preverbally in 11% of cases in Dohok. Table 4 also demonstrates that the deviant position of P is slightly more common in Dohok (i.e., PV instead of VP), suggesting a greater flexibility of objects in this dialect ($p<0.0001$).

Table 4: Total A and P placement variants.

	Dohok	Sanandaj		Dohok	Sanandaj
AV	170/186	208/215	VP	254/286	18/411 (4.4%)
VA	16/186 (8.6%)	7/215 (3.3%)	PV	32/286 (11%)	393/411

A few of such postverbal Ps in Sanandaj occur in fixed idiomatic blessings (4/18); this presumably represents an archaic layer of the language when OV used to be more common:

(3)   Sanandaj   [Khan 2009: 412, A:17]
       ʔəlha   *šoq-Ø-la*   *ta-daăk-èf*
       god.MS   keep.IPFV-A.3MS-P.3FS   DOM-mother.FS-his
       'May God preserve his mother.'

Noticeable differences notwithstanding, some of the motivations for the deviating order are common to both dialects, and will thus be discussed together in Section 5.

4 The role of morphosyntax

Differential object marking (DOM) is expressed by differential object indexing (DOI) or flagging (DOF). Both Jewish varieties employ DOI, while DOF is only found in TZJ and is achieved through prepositional nominal marking (*əl-, ta-*). There is no direct correlation between word order and DOM strategies in Sanandaj. The fact, however, that DOF is a feature of the OV type is consistent with the cross-linguistic tendency to employ nominal marking when both A and O occur preverbally (Siewierska and Bakker 2009: 296–99).

In Dohok, there is a noteworthy correlation between DOI and word order in that DOI is significantly more likely for preverbal objects (p<0.0001), as shown in Table 5 below.

Table 5: DOI and O position in Dohok.

	OV		VO		Total (row)
DOI	27	(8)	46	(65)	73
no DOI	16	(35)	303	(284)	319
Total (column):		43		349	392

5 The role of information structure

Word order reflects the syntacticization of semantics and information structure (e.g., Comrie 1988). The most frequent permutations are generally grammaticalized for particular grammatical functions. They can also be grammaticalized for argument types by part of speech (cf. Section 5.2.).

In both dialects, the clause-initial position is typically associated with the topic role. This is a general feature of subject-initial languages (Herring and Paolillo 1995: 171). Clause-final position has been argued to be the preferred slot for focal arguments, yet this typically does not seem to hold for AOV languages (Herring and Paolillo 1995). Herring and Paolillo (1995: 193) emphasize that different types of focus should be taken into account. What constitutes a focus slot in a given language also correlates with frequency.

Cross-linguistically, in clauses with the pragmatically-unmarked broad focus (i.e., predicate-focus), the object belongs within the focus group, on the assumption that the object is a part of the verbal construction (cf. Lambrecht

1994: 296–297). The subject, by contrast, is topical. Thus, in a prototypical broad-focus clause such as *my sister hit a dog*, the subject *my sister* is the topic while the verb and the object *hit a dog* constitute the focus. In broad focus clauses in Dohok, stress (indicated by accentuation) typically falls on the clause-final constituent. This clause-final tendency is also found for Sanandaj, but is not complete in this dialect (see Section 5.3. below).

A change in an argument's position (along with the change of nuclear stress placement) signals that the pragmatic function of that argument is different from its prototypical one. This does not mean, however, that the default word order is always pragmatically unmarked. In both dialects, *wh*-focus is generally immediately preverbal or clause-initial. This incidentally partly coincides with the most frequent OV order in TZJ, but deviates from VO in LDJ. Similarly, the narrow focus (i.e., counter-presuppositional, contrastive, additive) generally appears immediately preverbal in TZJ dialects, as in other AOV languages like Korean and Armenian (Comrie 1988). The situation is more flexible in Dohok where, although narrow focus objects also tend to be preverbal, a few postverbal cases of narrow-focus object are also attested (Molin 2021: 385).

5.1 Topicalization through clause-initial position

In both dialects, any constituent can be placed clause-initially in order to establish it as a new discourse topic (cf. examples in 4 and 5a below) or to reactivate a discourse-old topic in cases where there is a switch in topic (5b). This is often but not always signaled prosodically by placing the NP in its own intonation unit (marked by "|" below). The constituent can be referred to inside the clause by a pronominal index, as illustrated in (5b). Thus, in both dialects, clause-initial placement is used for the topicalization of objects as well as Goals, as shown in (5b):

(4) Dohok [Molin fieldnotes]
 Sehrâne| g-oð-ax-wa...
 sehrane IND-do.IPFV-A.1PL-PST...
 '[As for] **Sehrane** – we used to celebrate [it] (after we finished Passover).

(5) Sanandaj [Khan 2009: 446, B:75]
 a. **găḷa-e ʔilanè**| bšəlmane čăq-èn-wa
  leaves-of trees Muslims pick.IPFV-A.3PL-PST
  '[As for] **leaves of trees** – the Muslims picked [them].'
 b. Context: 'In the evening he (i.e., the neighbor of the merchant) had to go home...'

> ***tajər-ăkè***   *ḥăsab-ef*   *haw-le-o*   *baq-èf*
> merchant-DEF account.MS-his give.IMP-T.3MS-back to-3MS
> '[As for] **the merchant** – he would give his accounts to him.' (436, A:106)

5.2 Arguments expressed by independent pronouns

Independent personal and demonstrative pronouns are usually mentioned early in the clause in both dialects, for example:

(6)   Dohok   [Molin field notes]
       ʔana   *b-qaṭəl-Ø-li*
       I   FUT-kill-A.3MS-P.1SG
       'He will kill **me**. (I won't come.)'

Only 4.6% (8/172) of pronouns occur postverbally in Sanandaj, and 3.3% (4/118) in Dohok. This preference is consistent with findings for subject-initial languages in general, in which the clause-initial position is associated with the topic role (cf. Siewierska 1998: 501). The referent of such pronouns is presupposed to be known to the interlocutors in the discourse, especially first/second person (speaker and addressee), and thus these pronouns are inherently topical. The favored initial position is regardless of the argument class (S/A/O) that they express, although independent pronominal Os like (6) above occur in only a handful of times in Dohok.

In the Dohok and Sanandaj corpora, their preverbal position can also be due to focalization. The narrow focus position tends to be immediately preverbal for pronominal objects in Sanandaj, for example:

(7)   Sanandaj   [W Iran; Khan 2009: 412, A:17]
       ʔana   ***did-i-č***   *qaṭl-i*
       I   OBJ-1SG-too   kill.IPFV-A.3PL
       '[As for] me, they will kill **me too**.'

5.3 Nominals in narrow focus

In Dohok, fronting the object and adding nuclear stress is a strategy for marking it for narrow focus (e.g., counter-presuppositional or additive focus). Consider the additive focus in (8) below; the predicate is out of focus, since the previous clause has already introduced the command 'to bring/buy':

(8) Dohok [Molin fieldnotes]
Context: 'Buy me seven jugs!'
ʔu-šoʔa ḥammàre žik muθ-un
and-seven donkey.drivers also bring.IMP-A.2PL
'And bring [me] **also seven donkey drivers**!'

Generally, the final constituent carries nuclear stress in broad focus clauses in both dialects, meaning that in Sanandaj, the stress typically falls on the verb or the Goal. However, nuclear stress in this dialect can also occur on the preverbal object if this object is discourse-new and indefinite (see below). Discourse-new Os appear in their usual preverbal position in Sanandaj (Khan 2009: 344–345) and typically carry nuclear stress, compare (9a) and (9b) below. Only in a handful of cases a postverbal O (with nuclear stress) expresses new information in Sanandaj (e.g., Khan 2009: 345). There is no indication that the postverbal position is used for narrow focus.[6]

(9) Sanandaj [Khan 2009]
Context: 'On New Year's day they would give a table (of gifts) to the groom.'
[TOPIC] [NEW O] [PREDICATE]
a. ga-ef **širìn** măt-i-wa
in-3MS sweets put.IPFV-A.3PL-PST
'They put **sweets** in it.' (422, A:55)
b. Context: 'The merchant went to him, brought the cloth and put it down for him.'
[TOPIC] [OLD O] [PREDICATE]
'ay-zíl-∅ jəns **ləb-∅-lè**
he-went.PFV-S:3MS cloth took.PFV-O.3MS-A.3MS
'He (the buyer) went (and) **took** the cloth.' (435, A:105)

In the Sanandaj dialect, the immediate preverbal position is generally also reserved for constituents in narrow focus. Thus, narrow focal Os in preverbal position almost always carry nuclear stress (cf. 10a). However, stress is not restricted to narrow focus objects; sometimes, it also occurs in discourse-new indefinite objects in broad focus clauses. This means that for Sanandaj one does not always find a formal distinction between objects in broad and narrow focus, and

---

6 According to Khan, discourse-new objects can be postverbal if they introduce constituents which 'become significant for the ensuing discourse' (Khan 2009: 345), that is, carry presentational focus. Such a function of postverbal objects is attested in some A/SOV languages (cf. Herring and Paolillo 1995: 191–192).

where such a distinction exists, it is only signaled by nuclear stress or its lack. This contrasts with Dohok, where broad and narrow focus objects are clearly differentiated by their position (VO vs. OV respectively).

In Sanandaj, the immediately preverbal position is also the dedicated narrow focus slot for other arguments such as A in (10b), G (10c) and oblique[7] (9d):

(10) Sanandaj [Khan 2009: 422, A:55]
 [TOPIC] [NARROW FOCUS] [PREDICATE]
 a. Context: 'Life was easy.'
 naše raba **tăqalà** la-dă-en-wa
 people.PL much exertion.MS NEG-put.IPFV-A.3PL-PST
 'People did not make much **strenuous effort** (i.e., exert themselves much).'
 b. ʔea **hulà-e** traste-ya
 DEM:SG Jew-PL made.FS-P.FS
 '**Jews** made this (i.e., not Muslims).' (470, B:83)
 c. ʔalha xa-tăra **baqa-didàn-ač** kol-∅-o.
 God a-door.MS for-1PL-too do.IPFV-A.3MS-back
 'God will open a door **also for us**.' (482, D:9)
 d. tanurake **ba-şìwe** malq-i-wa-la
 oven.FS.DEF with-wood.PL heat.IPFV-A.3PL-PST-P.3FS
 'They heated the oven **with wood**.' (424, A:67)

5.4 Discourse-old (topical) objects in preverbal position

As stated previously, the preverbal position is a slot for narrow focus in both dialects. At the same time, 61% of the preverbal objects in Dohok are discourse-old, indicating that the less frequent OV order also serves to defocalize O in Dohok (cf. Molin 2021: 382–384). Since Os generally occur preverbally in Sanandaj, the preverbal position is also the most frequent one for discourse-old Os. In both dialects, the lack of focus on the preverbal object and thus its topicality is typically signaled by the lack of nuclear stress on this constituent, meaning that absence or presence of nuclear stress on the object in the preverbal position tends to be pragmatically significant in both dialects. An example of

---

[7] Oblique arguments are defined here as prepositionally expressed arguments other than Goals (cf. Section 6).

stressed preverbal object for Dohok is given in (11) below; here, the object is topical and the predicate is in narrow focus:

(11) Dohok [Molin fieldnotes]
Context: A lorry came, a lorry car, full of Christians.
u-kullu **ʔan** **suraye** *qam-qaṭl-ĭ-lu.*
and-all DEM Christians PFV-kill-A.3PL-P.3PL
'And they killed **all of these Christians**.'

5.5 Tail-head linkage

A distinct phenomenon is tail-head linkage in which the speaker repeats the tail clause of the preceding paragraph completely or partially in the head clause of the new paragraph for discourse cohesion. This is a characteristic stylistic device in NENA narrative, illustrated in (12) for Sanandaj below.[8] Often, such repeated clauses are characterized by the inversion of the position of O relatively to the verb, as observed in (12b).

(12) Sanandaj [Khan 2009: 439, B:4]
a. **ga-mal-awae** *zəndəgì-kol-i-wa*
in-village.F-PL live-do.IPFV-A.3PL-PST
'(They went to the villages.) They lived **in the villages**.'
b. *ba-ʔaqle* **ʔay-jəle** *šuč-lu!*
with-feet DEM-clothes trample.IMP.A.SG-P.3PL
šuč-lu **ʔe-jəl-ăkè!**
trample.IMP.A.SG-P.3PL DEM-clothes-DEF
'Trample **the clothes** with your feet! Trample **the clothes**! (And I shall wring them out).' (476, C:11)

5.6 Postverbal S in Dohok

In Dohok, another difference between S and A in addition to frequency and type (nominal vs. personal pronoun, see Section 3.1.) concerns word order, as compared in Table 6 below. While A typically occurs preverbally (88%), 49% of nominal S arguments occur postverbally (Molin 2021: 403–424). In the case of A,

---

[8] For the various functions of this device in storytelling, see Coghill (2009: 277).

the postverbal position is largely lexically specific, since a single lexeme – ʔmr 'say' – accounts for more than a half of the VA cases.

Table 6: Position of full nominal S and A in Dohok.

	Postverbal		Preverbal		Total (row)
S	62	(41)	64	(85)	126
A	13	(34)	91	(70)	104
Total (column):	75		155		230

An exception to this tendency are clauses with Gs, which clearly prefer SVG order (30 against 4 cases of VSG). This, in turn, is apparently due to the above-mentioned preference for verb-medial position in clauses with two arguments (assuming that Goals of intransitive movement verbs are indeed argument-like).

The VS order in Dohok can largely be considered a marked structure for conveying sentence focus. Sentence-focal clauses (also labeled 'thetic' after Sasse 1987) contain no presupposed information, meaning that the entire sentence is in focus (Lambrecht 2000: 635). The statistically significant association ($p<0.0001$) with the postverbal position for discourse-new S is shown in Table 7 below:

Table 7: Position of full nominal S in Dohok by discourse status.

	Postverbal		Preverbal		Total (row)
Discourse-old S	62	(41)	64	(85)	126
Discourse-new S	13	(34)	91	(70)	104
Total (column):	75		155		230

Sentence-focal clauses are prototypically intransitive. The postverbal position of the Subject (typically of intransitive) in Dohok marks the constituent as a non-topic by placing it in the position which is associated with focal arguments (e.g., objects). This inversion of the Subject to code it is as a non-topic and thus to create sentence focus is characteristic of various other languages with the predominant AVO order (e.g., Italian, cf. Lambrecht 2000).

However, in the Dohok NENA dialect, this association is not perfect – the phi coefficient[9] for the data in Table 7 is 0.5, meaning that the relationship between discourse status and word order is moderate and the VS order likely also covers other pragmatic functions (see Molin 2021: 404). This demonstrates that word order does not always perfectly correlate with a functional pragmatic category (or in some cases covers various pragmatic functions). This point is important to make, since studies such as Lambrecht's (2000) can be taken to imply such a complete correlation between word order and pragmatics. Thus, in the Dohok dialect, discourse-new Subjects can occur preverbally and discourse-old ones can be postverbal. In some cases, the postverbal placement of a discourse-old Subject apparently occurs when a Subject is added as an 'afterthought' – often clarifying the topic in cases of topic shift. A strong association between discourse status and word order (p = 0.55) is found for inanimate Subjects. This means that discourse-new inanimate S arguments are especially likely to occur postverbally. Overall, the S argument in Dohok is more P-like than A-like in terms of its position, meaning that word order in this NENA dialect tends towards an ergative alignment.[10]

6 Obliques and ditransitive constructions

6.1 Goals vs. other obliques

Oblique arguments are defined here as prepositionally expressed arguments that do not belong to our category of Goals (see Section 2). These arguments can be further subdivided into: Locations, Sources and Instruments and other non-locative obliques.[11]

Stilo (2018) reports that in the NENA varieties found East of the Zagros Mountains, oblique arguments such as Instruments and Locations are generally preverbal, which contrast with the general postverbal position of these arguments elsewhere in NENA.

---

[9] The phi coefficient of 0 would mean that there is no relationship, while 1 would mean a perfect relationship between two variables (Stefanowitsch 2020: 153).
[10] A more detailed study of intransitive clauses in Dohok is included within Molin (2021). This study also gives the mathematical basis of the word order ergativity claim.
[11] These correspond respectively with Locative, Ablative, Instrumental and Comitative in Stilo (2018).

Prepositional complements typically follow the verb in both dialects we studied but not in the same rigor. The above findings conform to our study of the Sanandaj corpus and fit with the general profile of the TZJ dialects. Our study shows that 62% of obliques occur preverbally, while only 9% of Goals are preverbal (see further below). The percentages for the preverbal position of obliques for TZJ dialects in Stilo (2018), which he gives per individual category, have the mean average of 80.3%. As a representative of NENA in the West, Dohok, reveals an entirely different picture. About 97% of all prepositional arguments are postverbal. When both Source and Goal are expressed, Source precedes Goal. This preference for the postverbal position is thus even starker than for the Christian Barwar NENA dialect (North-West Iraq), about 80% in Stilo (2018).

If both dialects had the same VG typology, the expected frequency of preverbal Goals should be roughly the same. Putting clauses with Goals from both dialects together, the number of preverbal Goals is higher than expected in Sanandaj (43/493 observed against 28 expected) and lower for Dohok (2/301 observed against 18 expected). This demonstrates that the position of Goals is significantly more flexible in Sanandaj ($p<0.0001$).

When we, however, consider the position of Goals in light of other obliques in Sanandaj, the preverbal position of Goals is much lower than expected ($p<0.0001$), as shown in Table 8 below. Expected frequencies are given in parentheses.

Table 8: Observed and expected frequencies of Goals vs. other obliques in Sanandaj.

	Postverbal		Preverbal		Total (rows)
Goals	450	(390)	43	(103)	493
Other obliques	56	(116)	91	(31)	147
Total (columns)	506		134		640

Several prepositions are more likely to be postverbal than others. Only 3.8% (6 out of 157) of the Recipients, Beneficiaries and Addressees expressed by the preposition *baq-* 'to, for' occur preverbally. The preposition *reš* 'onto, upon' occurs only once (2.5%) preverbally against 40 times postverbally, 85% of which denote destinations of (caused) motion verbs and the remaining 12.5% denote locatives. Thus, Recipients, Addressees and the Goals of motion verbs show no significant differences. Finally, adverbials of motion verbs expressing the direction of motion such as *warya* 'outside' and *qame* 'forward' always follow the verb in the Sanandaj corpus. If they go before the verb, they express a location. By contrast, Source NPs marked by *mən-/m-* (also conveying 'with') are predominantly preverbal (92.7%, 76/82). Virtually all pronominal arguments expressed by this preposition follow the verb, however. Where both Source and Goal are expressed, the Source occurs preverbally earlier in the clause.

6.2 Ditransitives: the position of R and T

In the prepositional indirect object construction, the T-like object triggers DOI and the R is prepositional (and thus subsumed under Goals here). As expected, LDJ dialects such as the one spoken in Dohok show different tendencies for the linear order of postverbal objects in prepositional indirect object constructions based on the type of the argument (by part of speech etc.). Prepositional pronouns tend to immediately follow the verb.

(13)   Dohok
a. ʔaxni   məθe-lan   **ṭa-lox**   diyàr-e
    we   bring.PFV-A.1PL   to-you.MS   gift-PL
    'We brought **you** presents.'
b. zun-nu   diyàr-e   **ta-bab-i**
    buy.PFV-A.3PL   gift-PL   for-father.MS-my
    'They bought presents **for my father**.'

As in the majority of languages (Malchukov et al. 2010), double object constructions in NENA are lexically restricted (cf. Khan 2008a: 785). These ditransitive verbs have the causal profile 'A causes R to receive T' (Blansitt 1984) such as *mly* 'fill R T'. The first (or primary) object typically outranks the second (or secondary) one in affectedness and topicality and is considered the most salient affectee much like a patient (e.g., Givón 1984). In these constructions, the R-like object triggers differential indexing like P. An exception to this are a few verbs like verbs of 'making T into R'. The position of R and T is pragmatically determined: the topical argument is usually preverbal (or occasionally immediately postverbal).

Interestingly, in both Sanandaj and Dohok, the more topical R tends to be mentioned early onwards preverbally. This preverbal preference is characteristic of topical arguments in general, cf. Section 5.1 and Section 5.2. Thus, the R-like argument in verbs of naming is always (4 cases) preverbal due to its topical status in Dohok, as in (14).

(14) Dohok [Molin fieldnotes]
    Dŏhok   k-ṣarx-í-wa-la   màθa.
    Dohok   IND-call.IPFV-A.3PL-PST-3FS   town.FS
    'Dohok used to be called (a) town.'

Verbs of naming are frequently attested in the Sanandaj corpus, as illustrated in (15a). The T always follows the verb in the double object constructions of naming. The name of the verb 'say' in the sense of 'calling R T' accounts for almost half of the total VO cases.

(15)   Sanandaj   [Khan 2009: 448, B:30]
   a.   *ʔoa-sawzì-jad*ˡ   k-əmr-i-wa   gilàxa
        DEM-mixed-herbs.MS   IND-say.IPFV.3PL-PST gilaxa
        'The mixed herbs they called *gilaxa*.'

The Ts of naming verbs are only preverbal when Rs are prepositional. Thus, an alternative to the clause in (15a) can be a clause with the TVR order (given in 15b). There, the Recipient is prepositional (and thus postverbal). Even when the R is prepositional, however, the secondary T can still comply with the postverbal tendency shown above, as illustrated in (15c), yielding the VRT order. In all cases, the R is topical and T focal.

(15) b. **šaplultà**   k-əmr-i-wa   baq-ef
         Šaplulta   IND-say.IPFV-3PL-PST   to-3MS
         'They called (lit. said to) it **šaplulta**.' (462.68)
   c.   k-əmr-əx-wa   baq-ef   tuḷèˡ   ʔilanè   xelapa
        IND-say.IPFV-1PL-PST to-3MS   shoot.PL   tree.PL-of   willow.MS
        'We called (lit. said to) them **shoots, of willow trees**.' (466, B:75)

In verbs of 'making T into R', by contrast, the more topical and most affected argument is the T, occupying the preverbal position, whereas R is postverbal. In Sanandaj, the object's position before or after the verb 'make' can be semantically significant. The postverbal position triggers a goal-like interpretation, compare (15d) and (15e). As an alternative to the clause in (15e), the Recipient can also occur as prepositional argument. One can therefore consider postverbal objects like *kăbāb* in (15e) to be like Goals in (15f).

(15)   [P]   [V]
   d.   **kăbā̀b**   kol-i-wa
        kebab.MS   make.IPFV-A.3PL-PST
        'They made **kebab**.' (448, B:35)

	[V]	[T]	[R]
e.	kol-i-wa	-le	**kăbâb**
	make.IPFV-A.3PL-PST	-T.3MS	kebab.MS

'They made it into **kebab**.' (448, B:35)

	[V]	[T]	[R]
f.	kol-i-wa	-le	**ba-lešà**
	make.IPFV-A.3PL-PST	-T.3MS	in-dough.MS

'They made it **into dough**.' (450, B:41)

6.3 Areal and diachronic factors

Stilo (2018) notes that oblique arguments tend to occur preverbally in the languages of the area. As shown by Stilo (2018) and this study, within NENA, this only applies to the dialects of the eastern periphery (which generally show a greater degree of convergence with Iranian). In case of these oblique arguments, a general preference for the preverbal position has been recorded in Kurdish and West Iranian languages.[12] This suggests that the preverbal position of obliques in NENA East of the Zagros Mountains is due to contact with Iranian languages.

For Kurdish, the postverbal arguments that show the strongest preference for the postverbal position fall within the semantic domain of 'endpoints' (Haig 2019: 34). In his study of 26 languages from 6 families, Stilo (2018), qualifying Frommer's (1981) hierarchy for spoken Persian, gives the following hierarchy showing a decreasing likelihood for arguments to occur postverbally:

(16)   Goal > Recipient > Beneficiary > Addressee > DO (def.) > DO (indef.)

This is consistent with our case study for Sanandaj in that indefinite Os are least likely to occur postverbally, besides definite Os. Goals of intransitive motion verbs like 'go' most likely occur postverbally. We did not observe statistically significant tendencies for the other roles (Recipient, Beneficiary, Addressee), however.

Apart from areal factors, another factor to consider is the iconicity principle (e.g., Givón 1985)[13], which presupposes that linguistic structures such as word order preferences show a one-to-one correspondence with human experience of

---

[12] See Stilo (2018) and Haig (to appear) for details in individual dialects/languages and further subcategorization of such arguments. Stilo (2018) labels all 'peripheral arguments'.
[13] We are grateful to the editors for making us aware of this literature.

events. This predicts that word order would reflect the conceptual order in the trajectory of events, so that Agents precede Patients. Iconicity has also been argued to be the governing principle in the syntax of Sources before Goals in Dutch and some Papuan languages (Schapper 2011). In such cases, arguments in the semantic domain of endpoints can be expected to occur towards the end of the utterance. The preference to express Sources before Goals in NENA confirms this tendency to map the temporal sequence of the movement 'from...to' onto the utterance, that is, to express end-points after the start-points (cf. Haig to appear).

This iconicity in the linearization of Agent-Patient and Source-Goal situations, however, does not make clear-cut predictions about the position of these arguments relative to the verb. Cross-linguistically, OV systems tend to prefer preverbal position of two objects in ditransitive constructions, i.e., TRV or RTV order (Haspelmath 2015). Hence, for a VO language such as NENA dialects like Dohok, we expect both objects to be postverbal, for OV like Sanandaj we would expect both to be preverbal. Yet, we find the typologically unexpected combination of TVR and RVT order for ditransitives with either the VO or OV type. Moreover, the present study demonstrates that such typological generalizations require nuance through recourse to lexical restrictions and pragmatic roles. The preverbal position is the typical slot for the more topical argument, regardless of its argument class (T, R) and regardless of the type of ditransitive construction (prepositional indirect, double object), unless one of the arguments is focalized to preverbal position. Pragmatic functions of the argument, therefore, have to be considered in typological generalizations. Still, the preferred postverbal position for the semantically R-like argument arguably fits with the areal profile of reserving this position for the 'endpoint' roles.

Finally, the above factors notwithstanding, we briefly mention that diachronic factors should also be investigated. Givón (1979: 254), for instance, argues for a VO-to-OV continuum that allows for an intermediate stage where the direct object is preverbal but the indirect object lacks behind in occupying preverbal position. A diachronic study of preverbal Goals in well-documented pre-modern Aramaic languages such as Syriac could shed further light on NENA word order preferences.

7 Conclusion

In this paper we examined the word order preferences of objects and prepositional arguments in intransitive, monotransitive and ditransitive clauses for two types of Jewish NENA dialects: Sanandaj representing OV order and Dohok VO. The relative frequency of argument types (NP, independent pronoun) per argument class (S, A, O, G) is about the same for both dialects. The respective constituent orders show differences for argument classes, in particular O (and S). Objects (as well as subjects) are more flexible in Dohok (11% OV cases). Preverbal objects are more likely to trigger differential object indexing. This contrasts with only 4.4% of postverbal objects in Sanandaj.

Nevertheless, there is also considerable overlap. Firstly, narrow-focus objects tend to be preverbal in both dialects. Secondly, both types of dialects share a postverbal preference for Goals that is characteristic of languages in the NENA-speaking area. In both dialects, prepositional pronouns tend to be immediately postverbal. Themes and Recipients are commonly placed on either side of the verb. Relative sequencing of Themes and Recipients in ditransitive constructions is governed by pragmatic factors in both dialects, the preverbal slot being preferred for topical arguments. The relative sequencing of Agent before Patients and Sources before Goals is presumably governed by the iconicity principle.

The VG-tendency, however, is stronger for Dohok in a statistically significant way, meaning that Goals as well as other obliques are more flexible in Sanandaj. Sources show a greater tendency to be preverbal in Sanandaj, alongside Instruments and Locations. Moreover, particular prepositions and adverbials show different frequencies, meaning that some Goals are more likely to be postverbal than others. In some cases, word order is the only clue to the clause structure in Sanandaj. Certain adverbs express a Location preverbally, but a Goal postverbally and ambivalent verbs such as 'make' express the Patient preverbally, but the Goal postverbally (i.e., in the sense of 'make T into R').

We also examined the tendency for subjects of the intransitive verb to occur postverbally, attested in half of the clauses with full nominal S arguments in Dohok. In Sanandaj, the VS order only occurs extremely infrequently. In Dohok, the VS order occurs with both inanimate and animate (agentive) Subjects. This order is associated in a statistically significant way with discourse-new arguments, and should thus be considered a marked structure for expressing sentence focus (clauses with all-new information). However, the association between VS and discourse-new Subjects, i.e., between word order and a pragmatic function, is far from being complete.

We believe that attention to such granular detail is key for the creation of any large-scale models for typology. Consider, for instance, the observation that narrow focus objects are preverbal in both dialects, and that in the VO dialect, the OV order correlates with definiteness. Such observations could well illuminate details of the diachronic shift in NENA from VO to OV, presumably under the influence of Iranian. Perhaps such a shift first became widespread for narrow-focal and/or topical objects, gradually extending to other types of objects (perhaps along semantics lines such as definiteness or animacy). Interestingly, Goals and ditransitive constructions are hardly affected by this shift. In order to establish this with greater precision, however, future detail-oriented research in this area is necessary.

Abbreviations

1, 2, 3	1st, 2nd, 3rd person
A	agent argument of a transitive clause
DEF	definite
DEM	demonstrative
DOF	differential object flagging
DOI	differential object indexing
DOM	differential object marking/marker
F	feminine
FS	feminine singular
FUT	future
IMP	imperative
IND	indicative
IPFV	imperfective
LDJ	*lishana deni* Jewish
MS	masculine singular
NEG	negator
NENA	North-Eastern Neo-Aramaic
O/OBJ	object
P	patient argument of a transitive clause
PFV	perfective
PL	plural
PST	past
S	subject of an intransitive clause
T	theme of a ditransitive clause
TZJ	Trans-Zab Jewish

References

Andrews, Avery D. 2007. The major functions of the NP. In Timothy Shopen (ed.), *Language typology and syntactic description*, vol. 1: *Clause structure*, 132–223. Cambridge: Cambridge University Press.
Blansitt, Edward L., Jr. 1984. Dechticaetiative and Dative. In Frans Plank (ed.), *Objects: Towards a theory of grammatical relations*, 127–151. New York, NY: Academic Press.
Coghill, Eleanor. 2009. Four versions of a Neo-Aramaic children's story. *ARAM Periodical* 21. 251–280.
Coghill, Eleanor. 2016. *The rise and fall of ergativity in Aramaic*. Oxford: Oxford University Press.
Comrie, Bernard. 1988. Topics, grammaticalized topics, and subjects. *Berkeley Linguistics Society* 14. 265–279.
Frommer, Paul. 1981. *Post-verbal phenomena in Colloquial Persian syntax*. Los Angeles California: University of Southern California PhD dissertation.
Givón, Talmy. 1979. *On understanding grammar*. Amsterdam & Philadelphia: John Benjamins.
Givón, Talmy. 1985. Iconicity, isomorphism, and non-arbitrary coding in syntax. In John Haiman (ed.), *Iconicity in syntax*, 187–220, Amsterdam & Philadelphia: John Benjamins.
Haig, Geoffrey. 2014. Verb-goal (VG) word order in Kurdish and Neo-Aramaic: Typological and areal considerations. In Lidia Napiorkowska & Geoffrey Khan (eds.), *Neo-Aramaic and its linguistic context*, 407–425. Piscataway: Gorgias.
Haig, Geoffrey. to appear. Post-predicate constituents in Kurdish. In Yaron Matras, Geoffrey Haig & Ergin Öpengin (eds.), *Structural and typological variation in the dialects of Kurdish*. London: Palgrave Macmillan.
Haig, Geoffrey & Stefan Schnell. 2016. The discourse basis of ergativity revisited. *Language* 92(3). 1–14.
Haspelmath, Martin. 2015. Ditransitive constructions. *Annual Review of Linguistics* 1. 19–41.
Herring, Susan C. & John C. Paolillo. 1995. Focus position in SOV languages. In Michael Noonan & Pamela Downing (eds.), *Word order in discourse*, 163–198. Amsterdam & Philadelphia: John Benjamins.
Hoberman, Robert D. 1989. *The syntax and semantics of verb morphology in Modern Aramaic: A Jewish dialect of Iraqi Kurdistan*. New Haven: American Oriental Society.
Khan, Geoffrey. 2008a. *The Neo-Aramaic dialect of Barwar* I–III. Leiden: Brill.
Khan, Geoffrey. 2008b. *The Jewish Neo-Aramaic dialect of Urmi*. Piscataway, NJ: Gorgias Press.
Khan, Geoffrey. 2009. *The Jewish Neo-Aramaic dialect of Sanandaj*. Piscataway, NJ: Gorgias Press.
Khan, Geoffrey. 2011. North-Eastern Neo-Aramaic. In Stefan Weniger (ed.), *The Semitic languages: An international handbook*, 707–724. Berlin & Boston: De Gruyter Mouton.
Lambrecht, Knud. 1994. *Information structure and sentence form*. Cambridge: Cambridge University Press.
Lambrecht, Knud. 2000. When subjects behave like objects. *Studies in Language* 24(3). 611–682.
Levshina, Natalia. 2015. *How to do linguistics with R*. Amsterdam & Philadelphia: John Benjamins.
Malchukov, Andrej, Martin Haspelmath & Bernard Comrie (eds.). 2010. *Studies in ditransitive constructions: A comparative handbook*. Berlin & Boston: De Gruyter Mouton.

Meehan, Charles & Jacqueline Alon. 1979. "The boy whose tunic stuck to him": A folktale in the Jewish Neo-Aramaic dialect of Zakho. *Israel Oriental Studies* 9. 174–203.

Molin, Dorota. 2021. *The Jewish Neo-Aramaic dialect of Dohok: A comparative grammar*. Cambridge: University of Cambridge Ph.D. Thesis.

Mutzafi, Hezy. 2008. T*he Jewish Neo-Aramaic dialect of Betanure (Province of Dihok)*. Wiesbaden: Harrassowitz.

Noorlander, Paul M. 2014. Diversity in convergence: Kurdish and Aramaic variation entangled. *Journal of Kurdish Studies* 2. 202–224.

Noorlander, Paul M. 2018. *Alignment in Eastern Neo-Aramaic languages from a typological perspective*. Leiden: Leiden University Ph.D. Thesis.

Noorlander, Paul M. 2021. *Ergativity and other alignment types in Neo-Aramaic. Investigating morphosyntactic microvariation*. Leiden: Brill.

Sasse, Hans-Jürgen. 1987. The thetic categorical distinction revisited. *Linguistics* 25. 511–580.

Schapper, Antoinette. 2011. Iconicity of sequence in source and goal encoding in two Papuan languages of south-east Indonesia. *Linguistics in the Netherlands* 28. 99–111

Siewierska, Anna. 1998. Variation in major constituent order: A global and a European perspective. In Anna Siewierska (ed.), *Constituent order in the language of Europe*, 475–552. Berlin & New York: De Gruyter.

Siewierska Anna & Dik Bakker. 2009. Three takes on grammatical relations: A view from the languages of Europe and North and Central Asia. In Pirkko Suihkonen, Bernard Comrie & Valery Solovyev, (eds.), *Argument structure and grammatical relations: A crosslinguistic typology*, 295–324. Amsterdam & Philadelphia: John Benjamins.

Stefanowitsch, Anatol. 2020 *Corpus linguistics: A guide to the methodolog*y. Berlin: Language Science Press.

Stilo, Donald. 2018. Preverbal and postverbal peripheral arguments in the Araxes-Iran linguistic area. Invited Lecture at the Conference Anatolia-Caucasus-Iran: Ethnic and Linguistic Contacts, Yerevan University, 10–12 May 2018. available online: https://www.uni-bamberg.de/fileadmin/aspra/05_Events/2019_Post-predicate_elements_in_Iranian/2019_stilo_2018_yerevan_wordorder.pdf.

Index of Authors

Abdrahmanov, Makhmud A. 184f., 191
Akkuş, Faruk 229
Alexiadou, Artemis 153
Allan, Keith 21
Alon, Jacqueline 236, 258
Anagnstopoulou, Elena 153
Anatolian Arabic 198f., 202, 211
Anderson, Gregory D. S. 184
Andreas, Friedrich Carl 58
Andrews, Avery D. 238, 257
Anklesaria, Behramgore Tahmuras 60
Anonby, Erik 163
Arkadiev, Peter 129
Arnold, Jennifer 11, 21
Asadpour, Hiwa 39, 44, 64, 75, 90, 127, 165, 192, 218f.

Bakker, Dik 242, 258
Barbati, Chiara 57
Barjasteh Delforooz, Behrooz 94, 96, 115f., 122
Bashir, Elena 95f.
Beaver, David 143
Behaghel, Otto 2, 4ff., 78, 80
Behnstedt, Peter 199, 203
Belikova, Olga B. 182f.
Benveniste, Émile 40, 43, 45ff., 50ff.
Biberauer, Theresa 66, 156f.
Biryukovich, Rimma M. 184f.
Bittner, Maximilian 167
Blanc, Haim 199ff., 205
Bloch, Ariel A. 225
Bochi, Giovanni 127
Bögel, Tina 79
Bolinger, Dwight 5
Boni, Roza A. 184f.
Boyajian, Vahe 93
Boyarshinova, Zoya Ya. 182
Boyce, Mary 57f.
Branigan, Holly P. 11f. Bresnan, Joan 10f.
Bulut, Christiane 163, 177f., 192
Büring, Daniel 143
Butt, Miriam 79

Camilleri, Maris 205
Campbell, Lyle 91
Cerruti, Massimo 128
Chafe, Wallace 17ff., 128
Chang, Franklin 7, 12
Chomsky, Noam 2, 141
Clark, Brady 143
Coghill, Eleanor 236, 247, 257
Comrie, Bernard 186, 242f., 257f.
Čunakova, Oljga Michajlovna 43ff., 49

DeLancey, Scott 20
DiGirolamo, Gregory J. 17
Dijk, Teun van 17f.
Dik, Simon 17
Dimitrakopoulou, Maria 151f.
Dryer, Matthew S. 16, 39, 90
Dulzon, Andrey P. 182ff., 188
Durkin-Meisterernst, Desmond 57

Ergin, Muharrem 164

Faghiri, Pegah 7f., 66
Fauconnier, Gilles 18f.
Firbas, Jan 17
Fischer, Wolfdietrich 205
Fisher, Ronald A. 73, 77
Frommer, Paul Robert 65f.
Futrell, Richard 8

Georgi, Johann G. 182
Gernsbacher, Morton 18
Gibson, Edward 7f., 78, 82
Gignoux, Philippe 54f.
Gildea, Daniel 8
Givón, Talmy 18, 21, 251, 253f., 257
Göksel, Aslı 131, 133, 137, 148
Goldenberg, Gideon 205f., 211
Goldin-Meadow, Susan 3f.
Green, Georgia 10
Greenberg, Joseph 2ff., 7, 78, 164
Guardiano, Cristina 156
Gundel, Jeanette K. 18, 143
Gündoğdu, Songül 66f.

https://doi.org/10.1515/9783110790368-012

Haddad, Youssef A. 226f.
Hahn, Michael 8
Haig, Geoffrey 40, 59, 64ff., 89ff., 96, 122, 127, 137, 236, 239, 253f., 257
Haiman, John 78, 90
Hammarström, Harald 3
Harris, Alice C. 91
Harrison, K. David 184
Haspelmath, Martin 131, 254, 257
Hasselbach-Andree, Rebecca 219
Hawkins, John A. 2, 5ff., 10, 66, 78f., 82
Henning, Walter Bruno 58
Herin, Bruno 128
Herring, Susan C. 242, 245, 257
Hoberman, Robert D. 236, 257
Hockett, Charles 17
Holes, Clive 201, 213
Hollis, Alfred Claud 21
Holmberg, Anders 137

Iggesen, Oliver A. 129
Ingham, Bruce 199, 202f.

Jacobs, Joachim 17f.
Jahani, Carina 66, 90ff., 94, 114f., 122
Jastrow, Otto 199ff., 212ff.
Johanson, Lars 164, 178
Jügel, Thomas 39f., 48, 50, 58f., 66, 71, 73, 76, 117, 119, 122, 218f.

Kaltsa, Maria 152
Kayne, Richard 2
Keenan, Edward 3f.
Kemp, Charles 78
Kerslake, Celia 131, 133, 137
Khan, Geoffrey
Khan, Geoffrey Allan 40, 201, 205, 211f., 219, 222, 229f., 235ff., 243ff., 251f., 257
Kirk, Alison 144
Koopman, Hilda 21
Korn, Agnes 66, 73, 92ff., 121
Korta, Kepa 15
Kostrov, N. A. 183
Kotwal, Firoze M. 40
Krifka, Manfred 143

Lambrecht, Knud 17f., 20, 51, 143, 242, 248f., 257
Leitner, Bettina 203
Lemskaya, Valeriya M. 181, 184ff., 188, 191
Levshina, Natalia 238, 257
Lewis, M. Paul 16
Liu, Haitao 8
Lohse, Barbara 9f.

Malchukov, Andrej 251, 257
Malov, Sergey E. 185, 189f.
Masica, Colin P. 129
Mastropavlou, Maria 151
Matić, Dejan 17
Matras, Yaron 128f.
Medeiros, David J. 9
Meehan, Charles 236, 258
Megerdoomian, Karine 67
Melnick, Robin 9
Mithun, Marianne 67
Molin, Dorota 237, 243ff., 249, 252, 258
Molnár, Valéria 143
Mpaayei, John T. 21
Myachykov, Andriy 17f.

Neocleous, Nicolaos 141ff., 151f.
Nöldeke, Theodor 230
Noorlander, Paul M. 235f., 238f., 258
Nourzaei, Maryam 93ff., 117, 119

O'Shannessy, Carmel 128
Oehrle, Richard T. 10
Öpengin, Ergin 66f., 70f., 77
ʿOryān, Saʿīd 48
Öztürk, Balkız 148

Palva, Heikki 198, 200ff., 205
Paolillo, John C. 242, 245, 257
Payne, Doris L. 15f., 21f., 28, 35, 59
Perry, John 15
Pöchtrager, Markus A. 148
Pomorska, Marzanna 184
Ponaryadov, Vadim V. 192
Posner, Michael I. 17
Prince, Ellen 18
Procházka, Stephan 198

Radloff, Wilhelm 182ff., 187
Reckendorf, Hermann 211
Regis, Riccardo 128
Reinhart, Tanja 17
Roberts, Ian 156f.
Rooth, Mats 17, 143

Sadler, Louisa 205
Samvelian, Pollet 7f., 44, 66f., 69, 71
Sardinha, Katie 128
Sasse, Hans-Jürgen 248, 258
Schapper, Antoinette 254, 258
Schmitt, Rüdiger 39
Schnell, Stefan 239, 257
Schönig, Claus 181
Schreiber, Laurentia 141
Şener, Serkan 153
Sidwell, Paul 75
Siewierska, Anna 240, 242, 244, 258
Simpson, Jane 1
Sims-Williams, Nicholas 43, 46f.
Sitaridou, Ioanna 141f., 146, 149, 152, 156f.
Skjærvø, Prods Oktor 39
Song, Jae Jung 2
Stalnaker, Robert 15, 18, 20
Stefanowitsch, Anatol 249, 258
Stilo, Donald 66, 236, 249f., 253, 258

Strahlenberg, Philipp J. von 183
Sundermann, Werner 43, 47

Tafazzoli, Ahmad 54
Talay, Shabo 199ff., 205
Temperley, David 8
Tokmashev, Denis M. 185, 192
Tomlin, Russell 11, 17
Tsimpli, Ianthi Maria 151f.
Tucker, Archibald N. 21

Valldoví, Enric 17
Vallejos, Rosa 17
Versteegh, Kees 198f., 202
Vilkuna, Maria 17

Wal, Jenekke van der 143
Walter, Mary Ann 205
Wasow, Thomas 4f., 10, 59, 78, 178
Watson, Janet C. E. 199, 201
Watters, John 17
Wedgwood, Daniel 17

Yamashita, Hiroko 7, 12

Zemer, Hila 205
Zúñiga, Fernando 16, 20

Index of Languages

Akkadian 219
Arabic 127f., 131f., 134f., 137f., 167, 178, 197ff., 202ff., 214, 224, 227, 229f., 235
– Anatolian ~ 198f., 202, 211
– Āzəx 203, 209, 212, 219, 226
– Bahraini ~ 213
– Bəḥzāni 212, 217, 227
– Betanure 237
– Classical ~ 205f., 210f., 221f.
– Dēr iz-Zōr ~ 204, 207, 209, 222
– Iraqi ~ 198f.
– Ka'bīye 221ff., 227f.
– Kəndērib 207, 209f., 212, 225
– Khuzestan ~ 199
– Kurdistan ~ 202, 212, 230
– Mardin ~ 203, 207, 209f., 213ff., 218ff., 227ff.
– Maronite Cypriot ~ 205
– Mesopotamian ~ 197ff., 202, 206, 213, 229f.
– Old ~ 199f.
– Qur'anic ~ 210f.
– Qarṭmīn ~ 210, 212f., 215ff., 226, 228ff.
– Uzbekistan ~ 200, 205
Aramaic 57, 123, 229, 231, 235ff., 254, 257f.
– Amedia 236
– Neo-Aramaic 205, 235f., 257f.
– Syriac ~ 56
Armenian 65, 67, 235, 243
Avestan 56
Azeri (Turkic) 65, 67, 198

Bactrian 39–59
Balochi 90ff., 103, 108f., 114ff., 121ff.

Chulym Turkic 181ff., 189ff.
– Čibi 182
– Ketsik 182
– Küärik 182, 185

Domari 127ff., 136ff.

English 5, 9, 11

Georgian 148
German 105, 111
Gilaki 93
Greek 141f., 144ff., 148ff., 157
– Cappadocian ~ 146, 149f.
– Phárasiot ~ 149f.
– Pontic ~ 146, 149f.

Iranian/Iranic 39–60, 163f., 166ff., 177f.

Japanese 7

Ket 183
Khakas 181, 183
Korean 7, 243
Koroshi 93, 114f., 117
Kurdish 64ff., 89ff., 93, 117, 123, 177, 198, 219, 222, 229, 231, 235, 253, 257f.
– Northern ~/Kurmanji 40
– Northeastern ~ (NEK) 65
– Mukri ~ 63ff., 70, 75, 79, 82f.

Ladino 202
Laz 148

Maa 16, 18, 21, 35
Melet 182, 184f., 189
Melets 185
Mḥallami 203, 209f., 222

North-Eastern Neo-Aramaic 91, 205, 212, 219, 222, 229, 231, 235, 256
– Arbəl 209, 223, 226
– Christian (Neo-)Aramaic 200ff.
– Christian Baghdadi 204, 230
– Christian Qaraqosh 230
– Dohok 235ff., 239ff., 254f., 258
– Jewish (Neo-)Aramaic 202
– Jewish Baghdadi 230
– Jewish Berber 202
– Sanandaj 235ff., 250ff., 257

Parthian 39–59
Persian 58, 65, 69, 71, 90ff., 115, 166f., 169, 177ff., 235
– Colloquial ~ 117, 123, 177
– Middle ~ 39–59
– Modern ~ 177ff.
– Old ~ 121
– Spoken ~ 253
Pumpokol 183

Rīš 222, 225
Romeyka 141–157
Russian 181, 183f., 188, 190ff.

Samoyed 183
Shor 181, 183
Siirt 203, 206ff., 221f., 225
Sílliot 149f.
Sogdian 40–59

Syriac 56f., 205, 230, 254 (see Arabic and Aramaic

Tajik 90f.
Tofa 181
Turkish 127f., 131, 133f., 137ff., 141f., 145ff., 163, 165, 198, 229
– Karamanlidika ~ 146f.
Turkoman 198
Tutal 182, 184f., 189, 192
Tuvan 181

Yači 182
Yeži 182, 185
Yiddish 202
Yugh 183

Zakho 235ff., 258
Zoroastrian Dari 202

Index of Subjects

adpositions
adposition 63f., 72, 95, 101, 117, 121
adpositional phrases 67, 90
circumpositions 76, 79, 164
cliticized forms 64
postposition 4, 6f., 57, 78f., 91f., 95, 105, 121, 164, 166ff., 175
preposition 4, 6f., 10f., 19, 23, 40, 44, 57f., 68ff., 72f., 75f., 78f., 82, 90, 92, 95, 104, 117, 121, 131f., 134, 137f., 166ff., 177, 202, 207, 210f., 218ff., 238ff., 242, 246, 249ff.
prepositional phrases 6, 79, 137f., 164, 166f., 177

alignment
accusative construction 45
alignment 167, 171, 230, 236, 238, 249
ergative construction 45f., 54, 58, 70

animacy
Animate First Principle 34 (*see also* **word order principles**)
animacy/animate 11f., 15, 24, 29ff., 63, 74, 76f., 83, 151f., 256
human 2, 8, 12, 17, 20, 22f., 34f., 95–104, 115ff., 121, 253
inanimate 11, 20, 24, 30, 33f., 76f., 121, 210, 217, 249, 255
local reference to humans 95, 121
non-human 97, 100ff., 115ff.

areality *see* contact

arguments (*see also* **object, subject,** and **Target**)
ablative expressions 112, 118
concomitance argument 111
final state arguments 66
non-prepositional recipient-like argument 127, 134, 238
non-subject core argument 16
patient-like complement 238
peripheral arguments 66, 253
postverbal argument 89ff., 110, 118, 120, 253
prepositional complement 44, 250
purpose-like argument 110, 118

case
ablative case 112, 118, 130f., 134, 165, 249
case marking 6, 63, 75, 127, 129, 164ff., 177
dative alternation *see* **word order**
nominative case 21, 94, 164
locative case 24, 95, 118, 121, 165, 249f.
object case 94
oblique case 45f., 75, 92, 94f., 110, 117f., 121, 129ff., 177

clause types
adverbial action clause 164
complement clause 48f., 56, 164
complex sentence 164, 167, 169f.
ditransitive clause 16, 21, 27ff., 239, 255
infinitive clause 51
intransitive clause 238, 249
main clauses 19, 42, 55, 70, 121, 142ff, 148f., 157, 165, 167ff.
matrix clause 164, 166
monotransitive clauses 16, 28, 255
relative clause 43ff., 58, 70, 81f., 164, 168, 210f.
subordinate clause 42, 96, 121, 141f., 150ff.
transitive clauses 2f., 11

cognition
cognitive closeness 20
cognitive foundation 18
cognitive preference 4 (*see also* **word order**)
cognitive proximity 20
cognitive world 19
cognitively activated 20 (*see also* **information structure**)
comprehender's interpretation 19
familiar concepts 18, 20 (*see also* **definiteness**)
framing element 19
human cognition 2

Index of Subjects

identifiability (*see* **information structure**)
literal space 20
mental space 18f.
physical proximity 20
processing 2, 5ff., 11, 18f., 29, 66, 78, 83
remoteness 20
semantic association 4
short-term memory 7f., 17f.
interpretable 151ff.
uninterpretable 151ff.
see also **iconicity**

communication
communicative actions 15
communicative functions 1
communicative efficiency 1, 8
communicative intentions 15

contact
bilingualism 128, 183
calques 56, 188, 190f.
contact-induced syntactic change 156f., 178f.
contact zone 90ff., 96, 122, 181, 198
convergence 198, 253
language contact 59, 90f., 155ff., 181, 192, 198f., 229

copula
copulas 197f., 205f., 208, 213f., 229f.
demonstrative copula 211ff., 217
negative copula 209, 216
positive copula 206f., 209f., 214f., 217, 229
relative copula 210f., 213
zero-copula 205, 207

data (*see also* **genre & style** and **methods & theories**)
corpus data 6, 10, 239
crowdsourced data 67

definiteness
definite 34, 66, 94, 164, 202, 204f., 220, 253
indefinite 66, 165, 199, 245, 253
referential 22, 24, 30

deixis
distal 16, 20f., 35
itive 20

proximal 16, 20f., 35
ventive 20

discourse (*see* **information structure**)

ezâfe
ezâfe 64, 67ff., 73f., 77, 79, 82f., 166f.

flagging (*see also* **adposition, case, markedness** and **marking**)
flagged/flagging 40, 48, 63f. 67ff., 73–83, 166ff., 242

focus
additive 135, 243f
argument focus 17
contrastive focus 17, 53, 73f., 143–150, 224, 228
counter-presuppositional 243f.,
exclusive focus 17
focal arguments 242, 248
focus strategies 42, 53
narrow focus 243ff., 255
non-contrastive focus 17
non-focused Targets 73
predicate(-centered) focus 17, 242
polar focus 17
restricting focus 17
sentence-focal clauses 248, 255
thetic 19, 248

genre & style (*see also* **methods & theories**)
genre 56, 184f.
official register 165
spoken language 1, 237
style 42, 51, 56ff., 115

grammaticality
acceptability judgments 9
disharmonic 40
ungrammatical(ity) 1, 57, 72, 81f., 121
well-formedness 1
harmonic word order (*see* **word order**)

iconicity
iconic(ity) 21, 59, 66, 76, 90, 100, 112, 118, 137, 174, 177, 253ff.

incorporation
incorporated position 44, 64, 66–69, 72ff
incorporated Targets 42, 63f, 66f., 73ff.
incorporation 67, 74, 79, 82, 219, 230

information structure (*see also* **focus** and **topic**)
activated status 17
anaphoric continuity 30
asserted information 15, 19
attention 5, 9, 16f., 20f.
cataphoric continuity 18, 30
cognitive laser-like 17
cognitively salient 17
common ground 18, 20
counter-expectation 17
discourse coherence 3, 5, 10f.
discourse functions 3, 5
discourse-established 22f.
discourse-new 245, 248f., 255
discourse-old 243, 246, 248f.
discourse-pragmatics 16
discursive role 134, 136
familiar information 18
identifiable 21, 23, 28
information flow 15f.
information structure 10f., 15–21, 34f., 58f., 73, 84, 112, 120, 133, 138, 149, 176, 186, 191f., 198, 242
new information 10ff., 58f., 73, 107, 113, 115f., 120, 143, 165, 171, 175f., 245, 255
non-referential 22, 24, 30
old information 10ff., 18f., 58, 143
pragmatic interpretation 152
pragmatically determined 251
presupposed information 73, 248
qualitative assessment 29
referential 22ff., 29f., 45, 56, 227
scope 143

markedness
bare 64, 74ff., 83
default 100f., 107, 114ff., 120, 240, 243
marked 17, 21ff., 28, 34, 68, 73, 75, 105, 117, 121, 132–136, 143, 145, 164f., 170, 175, 177, 217, 220, 222, 226, 230, 237, 243, 248, 251, 255
pragmatically (un)marked 143ff., 242f.
unmarked 3, 21f., 89, 117, 131ff., 137, 143ff., 164f., 169, 173, 175ff., 228, 239

marking (*see also* **adposition, case** and **flagging**)
affixation 129
agglutinative structure 164
applicative 22f., 25ff., 32f.
differential object marking/DOM 130, 165, 241f.
direct object marking 94, 221
full/reduced flagging 64, 74, 77, 83
morphological marking 41, 164
prefixes 21, 28
superessive marking 131f.
Target-marking 127, 129, 131, 221
zero-marking 131

method & theories
addressation parameter 18
construction grammar 15
corpus-based approach 236
corpus studies 9
crowdsourcing grammaticality judgment 67
diachronic perspective 186, 189, 192
experimental protocols 19
experimental studies 9, 11
feature-based theory 141
frequent 4, 39, 78, 89f., 95–101, 105, 111f., 116–119, 121, 134, 176f., 199, 239, 241, 243, 246
– frequency 3, 73, 113f., 119, 122, 137, 166, 169, 172, 176, 236, 239, 242, 247, 250, 255
generative framework 2, 141
Greenberg's universal 3, 78
IC-to-word ratio 7
information-processing framework 19
postverbal order hypothesis 21
psycholinguistic 7
qualitative study 21
quantitative study 21
Syntactic Prediction Locality Theory 78
universal constraints 1
Universal Dependencies 8
universals of grammar 2 (*see also* **cognition**)

morphosyntax
agreement 6, 54, 70f., 157, 205f., 210, 238
– agreement marker 54
– verb agreement 152
compositional construction 68f., 71
compounding 129
fixed construction 68f.
modified 69
petrified 69, 83, 212
possessive construction 95, 166
quantified 30, 69
verbal complex 42, 44, 64, 67
verb-particle construction 9

object
direct object 9, 66, 69, 83, 89, 94, 107, 110, 112ff., 116, 118ff., 122, 131, 165, 171, 173, 175, 177, 187f., 191, 218f., 221f., 228ff., 254
indirect object 94, 113, 118, 165, 176f., 186, 188, 190, 192, 218, 224, 238, 251
differential object marking (*see* **marking**)
double-object construction 11, 238, 251f.
long object 55

part-of-speech (*see also* **adposition, copula** and **pronouns**)
PoS 73, 84, 239, 242, 251
bound morpheme 63, 67
definite article 202, 204, 220 (*see also* **definiteness**)
demonstrative 20, 117, 206f., 211ff., 217, 244
non-verbal elements 64, 67, 73, 79, 82
participles 164, 230, 236
presentatives 211ff.
relativizer 42, 44f., 49, 210f.
subordinator 44f., 121
temporal adverbs 20

pronouns
clitic pronoun 42, 45ff., 51, 54, 56, 63, 67, 70, 73, 95, 102
free pronoun 25, 27f., 34f.
independent pronoun 215, 219, 237, 239, 244, 255
orthotone pronoun 67, 70, 73
pro-drop 73, 89, 95f.

pronoun 20, 23, 33ff., 42, 45, 50, 53, 58, 60, 73, 89, 94, 117, 151f., 165, 170f., 202, 205ff., 226, 228ff., 239, 244, 247, 251

prosody
intensity contours 172, 174f.
low tone 22
nuclear stress 243ff.
pitch 73f., 133, 163, 169ff., 192
suprasegmental intonation 169, 171, 177
subordination by intonation 121

pragmatics
pragmatic inverse 21
presupposed 15, 18, 20, 34, 244, 248

semantic role
addressee 16, 22, 39f., 63, 65f., 90f., 95, 107, 109, 114, 116f., 127, 131, 135f., 198, 218, 226, 228, 238, 240, 244, 250, 253
agent 3f., 11, 21, 58, 94, 164, 254f.
allowee 22
benefactive 22f., 131, 134f.
beneficiary 66, 105ff., 114, 116, 118, 226f., 253
causee 22
destinations 63f., 250
enablee 22
endpoint 39, 64, 66, 127, 137, 253f.
experiencers 39, 71, 90
giver 21
Goal 15f., 21ff., 39f., 60, 65f., 90f., 95, 97ff., 104f., 107, 110, 112ff., 127, 131ff., 172, 176f., 220, 236, 238, 240, 243, 245f., 248ff.
instrument 21f., 25f., 32ff.
locations 10, 22, 95, 110f., 118, 121, 165, 249f., 255
malefactive 22f.
non-directional locations 118
patient 4, 238, 251, 254f.
recipient 10f., 16, 21f., 66, 105ff., 113f., 116ff., 127, 131, 134f., 222, 224, 227, 230, 238, 240, 252f.
reference point 20
semantic macro-role 35, 127
semantic roles 16, 63, 165, 191

source 21, 185, 250f., 254
theme 21ff., 34f., 135, 169, 176

subject
grammatical subject 3, 70, 190
logical subject 205
morphological subject 70

syntax
clause-combining strategies 167
frame 18
inter-clausal pivot 22
subclausal relations 164
subordination 121
syntactic dependency 5ff., 79
syntactic derivations 151
syntactic form 5, 16
syntactic-semantic interface 151
tail-head-linkage 96, 115, 120, 247

TAM
aktionsart 95
aspect 164
aspectual meaning 95
mood 164
resultative participle 236
tense 164
time 20, 22, 111

Target
Target/R 15f., 21–35, 39–51, 54–60, 63–83, 90, 96, 98, 109, 116–122, 127, 129–139, 165, 168, 171f., 175–178, 192, 197–200, 218–230, 238, 251–255
incorporated Targets *see* **incorporation**
non-focused Targets *see* **focus**
postverbal Targets 42, 64, 75f., 80, 118
preverbal Targets 42, 75f., 80f.
Target-marking *see* **marking**

topic (*see also* **information structure**)
aboutness 19, 34
afterthought 120, 175, 249
discourse topic 16f., 29, 31, 243
discourse-old topic 243, 246, 248f.
hanging-topic 70
macro-propositional topic 29f.
non-topic 248
participant topic 17, 29
sentence topic 17, 19
sequential topic 30f., 34
topic 16ff., 29ff., 34, 51f., 59, 70, 163, 169, 171, 173, 176, 192, 242–246, 249
topic agreement 70
topic continuity/continuation 21, 29ff., 35, 51
topic strategies 42, 51
topical status 21, 251
topicality 16, 20, 29, 33f., 51, 246, 251
topicalization 177, 216, 223, 226, 228, 243
unit topic 29, 31ff.

verb
auxiliary 54, 164, 174
complex predicate/verb 42, 44, 66f., 94, 118, 214, 217, 238
converb 164
finite verb 39, 41f., 58, 64, 73, 79, 82, 94, 164, 167ff., 218f., 222ff., 228
gerunds 164, 168f.
light verb 44, 94, 136
matrix verb 51
monotransitive 16, 26, 28, 141, 240, 255
multi-word verb/MWV 64, 67ff., 79, 81ff.
serial verbs 95, 119
three-place verb 63

verb types
caused-motion verbs 39, 63f., 127, 133
change-of-state verbs 64, 173
give verbs 63
inchoative verb 127, 136, 173f.
look verbs 64
motion verbs 39, 64, 97, 99ff., 104f., 107, 114, 116, 122, 131f., 250, 253
say/speech verbs 39, 63, 90f., 107, 114, 127, 131, 135, 228
show verbs 64
'turn into' verbs 109
vector verbs 95, 119
verba dicendi (*see* **say/speech verbs**)
verbs of transfer 39, 90f., 105, 114, 119

Index of Subjects — 269

word order
basic word order 2, 4, 8, 11, 39, 42, 49f., 79, 89, 236
canonical ordering 5, 9
dative alternation 9ff.
diachronic shift 256
dominant order 4
double object dative construction 10f., 238, 251f.
free word order 1, 223
final position 39, 50, 55, 137, 164, 169, 173, 191, 235, 240, 242f.
harmonic word order 4, 8
(immediate) postverbal position 21, 39, 42, 45, 51, 53ff., 59f., 64, 66, 69, 71–83, 89ff., 94, 97, 100f., 105, 107, 109ff., 120, 122, 133, 137f., 168–177, 239, 245, 248–255
initial position 42, 48, 51f., 56f., 80, 111, 166, 169, 171, 190f., 212, 242f., 244
mirror position 114, 116, 120
optimal ordering 78
placement rules 42, 58, 95
postverbal constituent 9, 59
postverbal NPs 21f.
postverbal phrases 35
preverbal position 41, 45, 64, 74ff., 79, 83f., 98–121, 136ff., 169, 171f., 177, 244ff., 250, 252ff.
verb-final 7, 114f., 131, 172, 240
verb-first 56, 240
verb-medial 240, 248
word order variation 1, 5, 59, 66, 133, 143, 152, 184, 186, 189, 191, 198, 223
see also **Target**

word order principles (*see also* **methods & theories**)
cognitive principles 74, 78
dependency length 7f., 79ff.
diachronic factors 253f.

Early Immediate Constituents (EIC) 6f., 9, 78, 80f., 84
geistig eng zusammengehörig 4, 6, 78
Gesetz der wachsenden Glieder 6
heavy NP shift 9
law of the growing constituents 6
long-before-short order 7f., 11
Minimize Domains (MiD) 5f., 78, 80f., 84
placement constraints 67
processing principles 78, 83
proximal before distal 21
resistance principle 156
semantic dependency 5, 9f.
semantic closeness 78, 80
short-before-long order 6, 8, 11f.
Syntactic Prediction Locality Theory (SPLT) 78ff., 84
weight 5–11, 15, 35, 117

word order typology
evolution of VO and OV 141ff., 150, 157
head-initial language 4, 8, 91
head-final language 4, 6ff., 11, 91, 168
left-branching 78, 92, 164, 166f.
noun-genitive order 90
OT 42
OV 4, 66, 78f., 90f., 122, 137f., 141ff., 236, 240ff., 246, 254ff
OVG/T 40, 46, 59
rigid word order 138, 164
right-branching 92, 166ff.
SOV 2, 4, 6, 39, 42, 89f., 138, 164, 169, 192, 245
subject-initial languages 242, 244
SVO 2, 4f., 21, 131, 149, 187, 192, 173
TO 42
VO 4, 66, 78, 138, 141ff., 236, 241ff., 246, 252, 254ff.
VSO 2, 4, 16, 21, 28, 35, 131

www.ingramcontent.com/pod-product-compliance
Lightning Source LLC
Chambersburg PA
CBHW060351190426
43201CB00044B/2012